BASIC QUANTUM MECHANICS

BASIC
QUANTUM
MECHANICS

Klaus Ziock

Associate Professor of Physics
University of Virginia

JOHN WILEY & SONS, INC.
NEW YORK · LONDON · SYDNEY · TORONTO

Library of Congress Catalog Card Number: 68-8954
SBN 471 98383X
Printed in the United States of America

PREFACE

This volume is the result of a one-semester course in quantum mechanics that I taught at the University of Virginia. The course was taken by physics majors in their senior year and by graduate students from other departments, mainly electrical engineering and astronomy.

The material presented somewhat exceeds what can be taught during one semester. When presented in class the chapter on scattering could obviously be left out. If scattering theory is desired, one could leave out Chapters 11 and 12.

The basic ideas, theorems, and techniques of quantum mechanics are developed along familiar lines. They are applied, whenever possible, to real physical systems which, at the level of this book, necessitates an emphasis on the physics of the hydrogenlike atoms.

A look at the table of contents shows that problems related to the hyperfine structure of hydrogen and positronium have been given much more emphasis than is customary in texts at this level. This has been done for the following reason: the hyperfine structure and its Zeeman effect offer a unique opportunity to demonstrate stationary and time-dependent perturbation theory and such abstract concepts as the mixing of states in actual physical systems of current research interest, yet with a minimum of mathematical difficulties. Matrices can be diagonalized exactly, summations usually run only over two values of the summation index, and orthogonality and normalization do not require integration over all space but are immediately obvious from the properties of the two dimensional state vectors involved.

I owe a debt of gratitude to Professor J. Eisenberg for numerous enlightening and enjoyable discussions.

KLAUS ZIOCK

CONTENTS

A GUIDE TO SYMBOLS, NOTATIONS, AND UNITS

In the choice of notations the author of a book is inevitably faced with a dilemma. On the one hand, it would be nice to keep the notation unambiguous; on the other hand, it is desirable to keep it conventional. An unambiguous notation would not only quickly use up all the known alphabets, it would also of necessity be unconventional. The letter m, for example, *is* commonly used to describe the mass of particles—predominantly of electrons—but it is *also* commonly used for the quantum number of the z-component of the angular momentum. In the case of such an obvious clash between convention and uniqueness this author has always sided with convention. To reduce confusion as much as possible the following table is offered. It does not claim to be complete since sometimes a variable or constant appears briefly only to be substituted into oblivion on one of the following pages. In such a case I feel that inclusion in this table would have introduced more confusion that it would have alleviated.

In general, **cgs units** have been used in this book. In the treatment of angular momentum, **atomic units** have been used in some places, to the extent of letting $\hbar = 1$, (see p. 99.)

1. MATHEMATICAL SYMBOLS

Symbol	Explanation or Definition	Introduced
$\mathbf{A}, \mathbf{B}, \mathbf{C}, \ldots$	Matrices	Appendix A.4
$\mathbf{a}, \mathbf{b}, \mathbf{c}, \ldots$	Vectors or their representatives	Appendix A.4
$\mathbf{r}$	The radius vector or its representative	
$A, B, C, \ldots$	Linear operators	Appendix A.1
$A, B, C, \ldots$	Scalars in general use. If (in a special case) $\mathbf{a}$ is	
$a, b, c, \ldots$	a vector, the symbol $a = |\mathbf{a}| = +\sqrt{(\mathbf{a} \cdot \mathbf{a})}$ is used for its magnitude	
i	The imaginary unit $i = \sqrt{-1}$	
$\mathbf{i}, \mathbf{j}, \mathbf{k}$	The unit vectors in the direction of the x-, y-, and z-axis of a cartesian coordinate system	
P_l	The Legendre polynomials	Eq. 5.47
P_l^m	The associated Legendre functions	Eq. 5.60
Y_{lm}	The spherical harmonics	Eq. 5.62
Δ	A finite increment of the quantity following Δ Example $a_2 - a_1 = \Delta a$	

1. MATHEMATICAL SYMBOLS (cont'd)

Symbol	Explanation or Definition	Introduced
∇	The gradient operator. In cartesian coordinates	

$$\nabla = \mathbf{i}\,\frac{\partial}{\partial x} + \mathbf{j}\,\frac{\partial}{\partial y} + \mathbf{k}\,\frac{\partial}{\partial z}$$

∇^2	The Laplace operator. In cartesian coordinates	

$$\nabla^2 = \frac{\partial^2}{\partial x^2} + \frac{\partial^2}{\partial y^2} + \frac{\partial^2}{\partial z^2}$$

	For the Laplace operator in spherical polar coordinates see Eq. 5.21	
$\sim$	As in $\tilde{\mathbf{A}}$ signifies the transpose of the matrix $\mathbf{A}$	Appendix A.4
†	As in $\mathbf{A}$† signifies the hermitian conjugate of the matrix $\mathbf{A}$	Appendix A.4
*	As in a^*, $\mathbf{a}^*$, λ^*, $\mathbf{A}^*$, etc., signifies the complex conjugate of a complex quantity.	
[ ]	Commutator brackets	Chapter 6.3

$$[\mathbf{A}, \mathbf{B}] = \mathbf{AB} - \mathbf{BA}*$$

2. PHYSICAL SYMBOLS

Symbol	Explanation or Definition	Introduced
c	The velocity of light	
D	Describing a state with $L = 2$	Chapter 7.5
d	Describing a state with $l = 2$	Chapter 5.6
d	The Lattice period of the one-dimensional crystal	Eq. 3.84
E	The energy	
e	The electronic charge, *also* the basis of the natural logarithm	
g	The gyromagnetic ratio	Eq. 7.14
$\mathbf{H}, H$	The hamiltonian matrix; the hamiltonian operator	Eq. 9.15; Eq. 2.30
$h, \hbar$	Planck's constant $h = 6.6256 \times 10^{-27}$ erg sec $\hbar = h/2\pi$; also in atomic units $\hbar = 1$	
I	The quantum number of the nuclear spin	
J, j	The quantum number of the total angular momentum of an atom	

* Also other uses

2. PHYSICAL SYMBOLS (cont'd)

Symbol	Explanation or Definition	Introduced
k	The wave number of an electromagnetic *or* matter wave $k = 2\pi/\lambda$, where λ is the wavelength	
L	The quantum number of the total orbital angular momentum	Eq. 5.114
l	The quantum number of the orbital angular momentum	Eq. 5.36
m	The quantum number of the z-component of an angular momentum, *also* the mass of a particle, especially the reduced mass (Eq. 5.8)	Eq. 5.29 Eq. 5.107 Eq. 5.8
n	The principal quantum number	Eq. 5.77
P	Describing a state with $L = 1$	Chapter 7.5
p	Describing a state with $l = 1$	Chapter 5.6
$\mathbf{p}, p$	The momentum, its magnitude	
$\mathbf{r}, r$	The radius vector, its magnitude	
S	Describing a state with $L = 0$	Chapter 7.5
s	Describing a state with $l = 0$	Chapter 5.6
t	The time	
$u(\mathbf{r}), u(r)$	A time-independent wave function	Eq. 2.18
V	The potential	
x, y, z	The components of the radius vector	
α	Spin state-vector indicating $m_I = \frac{1}{2}$	Eq. 10.65
	$$\alpha = \sqrt{\frac{2mE}{\hbar^2}} \; *$$	Eq. 3.4
β	Spin state-vector indicating $m_I = -\frac{1}{2}$	Eq. 10.66
λ	The wavelength*	
μ	A magnetic moment	
μ_0	The Bohr magneton	Eq. 7.11
ν	The frequency	
σ	The differential cross section	Eq. 13.1
σ_0	The total cross section	Eq. 13.2
$\sigma_x, \sigma_y, \sigma_z$	The Pauli matrices	Eq. 9.79
$d\tau$	The volume element, in cartesian coordinates: $d\tau = dx\, dy\, dz$, in spherical polar coordinates: $d\tau = r^2\, dr \sin\vartheta\, d\vartheta\, d\varphi$	
ψ	A (generally, time-dependent) wave function	
$d\Omega$	The solid angle element	Eq. 13.1
ω	The angular frequency	

* Also other uses

BASIC QUANTUM MECHANICS

INTRODUCTION

Since time immemorial, man has observed his environment and has tried to make sense out of what he saw. A crowning achievement in this endeavor was the creation of Newtonian mechanics. For the first time in history it had become possible to describe mathematically a large body of experience using just three basic laws. In the two centuries following Newton, physicists applied his theory in a more and more refined form to everything in sight. Buoyed by their brilliant successes, many physicists during this period believed that they would eventually be able to describe all natural phenomena in terms of Newtonian mechanics. Even the development of electrodynamics by Faraday and Maxwell seemed only to add new forces to the already known gravitational force, leaving Newton's laws untouched.

Toward the end of the last century and the beginning of the present one, cracks began to appear in the monolithic structure of physics. Experiments were performed whose results were in flagrant disagreement with any reasonable conclusion drawn from Maxwell's and Newton's theories. In this book we shall concern ourselves with these disagreements and with the conclusions drawn from them by the equals of Newton.

To physicists early in this century the failures of Newton's theory were deeply disturbing. To us, humbled by the struggle with the understanding of nuclear forces and, of course, equipped with 20/20 hindsight, its successes seem to be more startling than its failures. It is indeed almost miraculous that a theory which describes correctly the fall of the legendary apple on Sir Isaac's head also accounts for the motion of the earth with its 2×10^{25} apple masses around the sun. To expect that the same theory should also describe the motion of an electron with 3×10^{-30} apple masses seems now presumptuous.

When we enter the through-the-looking-glass-world of quantum mechanics, we must remember that our imagination has been molded in lifelong contact with things and events that are correctly described by Newtonian mechanics. It will, therefore, be best if we leave behind the collection of prejudices that

1

we sometimes fondly refer to as common sense. Again and again we shall have to examine with experiments the firmness of the ground on which we stand, and we shall have to entrust ourselves to the guidance of mathematics as we move about.

A BRIEF HISTORY OF QUANTUM MECHANICS

The great scientist, the true genius, blazes the trail into unexplored territory. For any serious student of science it will be inspiring and rewarding to trace the steps of the great explorers in their conquest of the unknown. Yet for his own first exploration, the student may find the steps of these explorers too steep and their trail too rough, and thus in this book our approach has generally been, *how it might have happened.*

The development of quantum mechanics, on the other hand, is one of the most fascinating chapters in the history of the human intellect; and in the following section we shall try to trace the historical development of the ideas presented in this book.

How It All Started

Toward the end of the nineteenth century, physicists had every reason to be satisfied with their accomplishments. Newtonian mechanics had explained the miracles of the heavens and had reached in its Lagrangian and Hamiltonian formulation an apex of mathematical elegance. Maxwell's equations had explained the mysteries of electromagnetism, and thermodynamics was a fully developed branch of physics and the secure foundation of a thriving technology. It is not surprising, then, that many physicists thought that all the questions has been asked and that finding the right answers would be merely a matter of time. One physicist who expressed himself in this sense was Phillipp v. Jolly. His remark would be just one of many famous last words if it were not for the name of the student to whom it was addressed. The student was **Max Planck** who, undeterred, had taken up the study of thermodynamics and by 1900, at the age of 42, had become one of the foremost authorities in this field. It was then that he presented at a meeting of the German Physical Society an *empirical formula* with which he attempted to bridge the gap between the Rayleigh Jeans law and the Wien law of blackbody radiation. The former described the connection between wavelength and intensity correctly at long wavelengths whereas the latter gave a correct description in the limit of short wavelengths. Planck's formula, which was a purely empirical interpolation between the two well known laws, fitted the precise measurements that were then available with extraordinary accuracy. This inspired Planck to search for a rigorous derivation of his formula[1] and

[1] M. Planck, *Ann. d. Phys.*, **4**, 553 (1900).

"...*after a few weeks of the most strenuous work of* [*his*] *life, the darkness lifted and an unexpected vista began to appear*..." This happened, however, not until he had been forced to "...*an act of desperation*."

This act of desperation was the assumption that an oscillator could absorb and emit energy only in the form of quanta of the energy $E = h\nu$. Planck's revolutionary assumption was ignored by most physicists and attacked by some. One of the most vigorous attackers was Planck himself, who for the following 15 years tried to derive his results without assuming the quantization of the oscillators. He came away knowing "...*for a fact that the elementary quantum of action played a far more significant part in physics than* [*he*] *had initially been inclined to suspect*...."

It was none other than **Albert Einstein** who realized in 1905 the sweeping significance of the assumption that Planck, who was no revolutionary, had made so reluctantly.

In 1905 Einstein[2] concluded that Planck's "...*determination of the quantum is to a certain degree independent of his theory of black body radiation*...." He then showed that Planck's "*light quantum hypothesis*" if generalized by assuming that all light can be emitted or absorbed only in the form of quanta of the energy

$$E = h\nu$$

explained not only Stoke's law of fluorescence but also Lenard's recent measurements of the photo effect. Einstein's equation $E = h\nu$ of course specifies only that light cannot be emitted continuously. It was not initially interpreted as meaning that *light quanta* are discrete particles that are emitted in a well defined *direction*. This final conclusion was drawn by Einstein in 1909.

Quantum Mechanics and the Atom

Today the terms atomic physics and quantum mechanics are almost synonymous, yet the application of the quantum hypothesis to the theory of atomic structure was slow in coming. The idea that atoms are the building blocks of all matter had been firmly established during the nineteenth century; however, the structure of the atoms remained a complete mystery. Without any notion at all of the atomic structure, it was of course impossible to apply the new quantum hypothesis to what we now consider its most proper realm. This situation changed suddenly in 1911 when an English physicist, **Ernest Rutherford**,[3] discovered that all the positive charge and almost all the mass of an atom are concentrated in an extremely small nucleus surrounded by an almost massless negative cloud. In 1912 a young Danish physicist, **Niels**

[2] A. Einstein, *Ann. d. Phys.*, **17**, 132 (1905).
[3] E. Rutherford, *Phil. Mag.*, **21**, 669 (1911).

Bohr, met Rutherford and one year later he had abstracted from Rutherford's discovery a theory of the structure of the hydrogen atom.[4] Bohr's model of the hydrogen atom had the electron circle the nucleus in *allowed* orbits whose *angular momenta* were quantized. The energy difference between two orbits was to be equal to the energy of the photon emitted in the transition from one orbit to the other.

During the next ten years or so Bohr's theory was generalized and refined and by 1923 it had been built into the complex system of postulates and empirical rules that is now known as the *old quantum theory*. This theory was capable of explaining most of the observed features of atomic spectra qualitatively and of explaining some of them quantitatively. All the while it was obvious that it was not the real thing.

Quantum Mechanics

The next step forward was an eerie hypothesis by **Prince Louis de Broglie**[5] of France. He proposed, based on relativistic considerations, that particles should be assigned a wavelength

$$\lambda = \frac{h}{p}$$

now known as the de Broglie wavelength. This sent an Austrian physicist, Erwin Schrödinger, who was at that time in Zürich, hurrying to his desk. But while Schrödinger still calculated, lightning struck in Göttingen, Germany, and set off an explosive development unequalled in the history of science. In July, 1925, **Werner Heisenberg,** one of many brilliant young men that had assembled in Göttingen under the tutelage of **Max Born,** sent to the Zeitschrift für Physik a paper[6] with the abstruse title "Über die quanten theoretische Umdeutung kinematischer und mechanischer Beziehungen." [7]

In this paper he proposed a quantum theory that did away with all such classical concepts as velocity and location of the electrons in an atom that, alas, could not be measured in any conceivable way and replaced them with relations between observable quantities. The algebraic rules that connected the observables, Heisenberg invented as he went along. In September of the same year his colleagues **Max Born** and **Pascual Jordan** pointed out[8] that Heisenberg's rules were the rules of matrix algebra, a mathematical subject that physicists in those days had little reason to study.

---

[4] N. Bohr, *Phil. Mag.*, **26,** 1 (1913).
[5] L. de Broglie, *Ann. de Physique*, **3,** 22 (1925).
[6] W. Heisenberg, *Zs. f. Phys.* **33,** 879 (1925).
[7] "On the quantum theoretical reinterpretation of kinematical and mechanical relations."
[8] M. Born and P. Jordan, *Zs. f. Phys.*, **34,** 858 (1925).

In the meantime **Erwin Schrödinger** had not been idle and in January, 1926, he sent a paper[9] to the "Annalen der Physik" with the title, "Quantisierung als Eigenwert Problem."[10] In this paper he introduced his postulates for the transition from classical mechanics to quantum mechanics, derived the Schrödinger equation of the hydrogen atom, and solved this equation in essentially the manner that has been presented in this book. One month later in February, 1926, in another paper,[11] he mentions that the perturbation theory of classical mechanics can be extended to quantum mechanics and concludes "...*In first approximation results the statement that the perturbation of the eigenvalue is equal to the perturbation term averaged over the unperturbed motion....*"[12]

In March, 1926, Schrödinger showed[13] the equivalence of his theory and Heisenberg's matrix mechanics.[14]

The theories that had thus far been developed were nonrelativistic. But the precision of spectroscopic measurements did not allow the small relativistic effects to be swept under the rug. The "fine structure" splitting had already been bothersome to the old quantum theory and in October, 1925, two Dutch physicists, **G. E. Uhlenbeck** and **S. Goudsmit,**[15] using the old quantum theory, explained this splitting as the consequence of an intrinsic angular momentum of the electron. They concluded that for this intrinsic moment the gyromagnetic ratio must be $g = 2$. In May, 1927, **Wolfgang Pauli**[16] was able to present a formal theory for this electron spin using matrix notation.

In the meantime, physicists all over the world had gone to work and had applied Heisenberg's and Schrödinger's ideas to the numerous problems of atomic physics that had been awaiting solutions. In January, 1928, **Paul Adrienne Maurice Dirac,** at Cambridge published[17] "The Quantum Theory of the Electron" which reconciled quantum mechanics with the special theory of relativity. He put the capstone on an intellectual edifice that in all its splendor had taken less than three years to be constructed.

[9] E. Schrödinger, *Ann. d. Phys.*, **79,** 361 (1926).
[10] "Quantization as an eigenvalue problem."
[11] E. Schrödinger, *Ann. d. Phys.*, **79,** 489 (1926).
[12] See Eq. 10.10.
[13] E. Schrödinger, *Ann. d. Phys.*, **79,** 734 (1926).
[14] See Chapter 9.
[15] G. E. Uhlenbeck and S. Goudsmit, *Naturwissensch,* **13,** 1953 (1925).
[16] W. Pauli, *Zs. f. Phys.*, **43,** 601 (1927).
[17] P.A.M. Dirac, *Proc. Roy. Soc.*, **117,** 610 (1928).

1

THE EXPERIMENTAL FOUNDATION
OF QUANTUM MECHANICS

In this Chapter we shall discuss a few experiments whose results do not agree with what classical physics would lead us to believe. The nature of the disagreement will make it obvious that a completely new theoretical approach is needed to describe these experiments. In selecting examples of these experiments, we shall not in general follow the historical development, but shall select those experiments that make the break most apparent. The inventors of quantum theory did not wait for most of these experiments to be performed. They were brilliant enough to read the new theory out of the less striking experimental evidence available to them.

1.1 THE PHOTOELECTRIC EFFECT

If light strikes a piece of metal it frees electrons from its surface, and it is experimentally possible to count these electrons and to measure their energy. A quantitative investigation of the effect shows that the light intensity determines the number of electrons thus freed but has no influence on their energy. This is contrary to what we would expect from Maxwell's electromagnetic theory since the intensity of a light wave is proportional to the square of the amplitude of its electric field vector. The energy of the electrons is, surprisingly, determined only by the color of the light, i.e., its frequency. **Albert Einstein** (1905) showed that the experimental results could be described by

$$E = h\nu - E_w \tag{1.1}$$

where E is the kinetic energy of the photoelectrons, h is **Planck's constant,**[1] ν is the frequency of the incident light, and E_w is the work function of the

[1] The constant $h = 6.6256 \times 10^{-27}$ erg sec was introduced in 1900 by **Max Planck** to describe the thermal radiation of a black body. We know today that h is a fundamental constant of nature which plays an all important and all pervading role in quantum mechanics.

6

metal, i.e., the amount of energy needed to remove an electron from the metal surface. This experiment seems to demolish at once the conventional notion that light is an electromagnetic wave. Einstein interpreted Eq. 1.1 by postulating that light always comes in the form of small packets, **light quanta** or **photons,** and that the amount of energy in each photon is

$$E = h\nu \tag{1.2}$$

In the photoelectric effect an electron absorbs a single photon whose energy becomes the kinetic energy of the electron.[2] The light intensity is simply given by the number of quanta per second. In other words, light, which we had thought consisted of electromagnetic waves, behaves in this experiment as if it consisted of individual particles with an energy $E = h\nu$.

We can derive the linear momentum of an individual photon by using Maxwell's electromagnetic theory.

A plane electromagnetic wave of total energy E_t transmits a linear momentum:

$$p_t = \frac{E_t}{c} \tag{1.3}$$

where c is the velocity of light. It follows that the momentum p of a single photon is given by

$$p = \frac{h\nu}{c} = \frac{h}{\lambda} \tag{1.4}$$

$$(\lambda = \text{wavelength})$$

This result is borne out by experiments. According[3] to Eqs. 1.2 and 1.4 it is not possible to transfer all the photon energy to the electron *and* to conserve momentum in a collision between a photon and an *unbound* electron. The photo effect requires that the electron is bound to an atom, whose recoil acts to conserve momentum. Since the atom is much heavier, the transfer of energy to the electron is almost complete.

The collision of a photon with a free, or loosely bound electron is known as the Compton effect. If a light beam[4] penetrates a thin slice of matter some of the light is scattered. **Arthur Compton** (1923) showed that the wavelength of the scattered light was increased and that the change in wavelength depended on the scattering angle. Compton was able to explain the measured angular dependence of the wavelength shift perfectly when he assumed that individual

---

[2] The energy E_w is due to surface phenomena. If the photoeffect is observed on free atoms, as in a gas, one has to replace E_w with E_i, the ionization energy of the atoms.
[3] See Problem 1.1.
[4] This experiment is usually done with light of a very short wavelength, i.e., with x-rays.

photons of energy $E = h\nu$ and momentum $p = h\nu/c$ collided with individual electrons in such a way that momentum and energy were conserved just as they are in the collision of a pair of billiard balls.

1.2 THE DIFFRACTION OF LIGHT

The fact that light behaves as if it consisted of particles even in experiments that also reveal its wave nature can be strikingly demonstrated in a simple diffraction experiment (Figure 1.1).

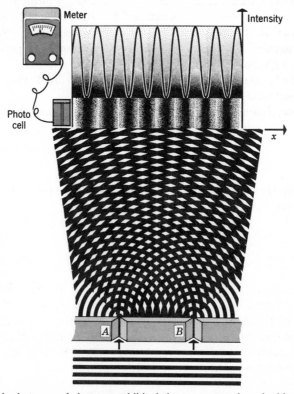

Fig. 1.1 Both photons and electrons exhibit their wave nature in a double slit diffraction experiment. They "interfere" with each other even if the intensity is so low that there is never more than one particle at a time between the slits and the screen.

A light beam penetrates the slits A and B and creates the familiar double slit diffraction pattern[5] on a screen. The light intensity on the screen is indicated in the diagram above the screen. If we move a photoelectric cell across the screen, its current will be proportional to the light intensity as shown in the diagram. Next we equip the photo cell with an amplifier that

[5] Shown here for slits that are narrow compared to the wavelength of the incident light.

enables us to register individual photoelectrons and thereby, according to Eq. 1.2, individual photons.

Now we put the photocell directly behind one of the two slits and reduce the light intensity until, according to the photocell, only every now and then a photon penetrates the slits. Having thus established that there is almost never more than one photon at a time between the slits and the screen, we return the photocell to the screen and move it slowly across. In doing so, we find that more photons arrive at the cell when it is at the location of a diffraction maximum and fewer, when it is at the location of a diffraction minimum. The number of photons counted during equal periods of time and plotted as a function of x gives exactly the same curve as the intensity plot obtained before. We also discover that at any location the *individual photons arrive at random*, although at a higher average rate at the diffraction maxima.

This experiment shatters all hope that we might be able to explain interference as some sort of interaction between photons that have penetrated the slits. Clearly a new theory is called for to reconcile the seemingly conflicting observations.

In sifting through the debris we find the following unassailable experimental facts:

1. The diffraction pattern predicted by Maxwell's theory exists even at the lowest light intensities.
2. If one slit is closed, a single slit diffraction pattern appears.
3. If monochromatic light is used, the photoelectrons in the photocell have all the same energy as given by Eq. 1.1.
4. At any given point "photons," as measured by the photocell, appear at random.
5. The *average* rate at which the photons appear is proportional to the light intensity.
6. In "collisions" of "photons" with electrons or atoms, energy and momentum are conserved according to the classical laws of physics.

From this we conclude, tentatively, that light consists of discrete particles—called photons—that can be counted. The photons obey the laws of a peculiar non-Newtonian mechanics. This mechanics seems to determine only where an individual photon is *likely* to go but does not seem to link cause and effect, i.e., initial and final conditions in the rigorous manner known from classical mechanics. The larger the number of photons involved in a measurement, the more closely their distribution approaches the distribution given by Maxwell's theory for the light intensity.

1.3 ELECTRON DIFFRACTION

We might console ourselves with the thought that photons are not really bona fide particles and that things will look different with, let us say, electrons.

So we replace[6] the lightbeam in Figure 1.1, with an electron beam and the photocell with some device capable of counting electrons. Again we measure the intensity, this time of the electrons as a function of the position x, and again we find a double-slit diffraction pattern. The electron "wavelength" calculated in the familiar manner from the slit separation, the slit to screen distance d and the diffraction pattern, turns out to depend on the electron momentum p. It is given by

$$\lambda = \frac{h}{p} \tag{1.5}$$

and is called the **de Broglie wavelength** of the electron.

It should be emphasized that this is the same relation that we had obtained for photons (Eq. 1.4). It was **de Broglie's** contribution to have guessed Eq. 1.5 correctly when there was no direct experimental proof for it.

The fact that Eq. 1.5 does not depend explicitly on the mass of the electron suggests strongly that it might hold true for any kind of particles, as, indeed, it does. We do not encounter this kind of diffraction phenomenon in our daily life simply because, as a result of the smallness of h, the wavelength of macroscopic bodies is exceedingly small. The wavelength of a man of $7 \times 10^4\,g$ walking at a velocity of 100 cm/sec is given by Eq. 1.5 as

$$\frac{6.6 \times 10^{-27}}{100 \times 7 \times 10^4} \approx 10^{-33}\ \text{cm}$$

Obviously this will not lead to observable diffraction phenomena if he passes through a slit of 100 cm width (a door).

1.4 PROPOSED THEORETICAL APPROACH TO THE PROBLEM

We have seen that light behaves as if it consisted of particles and that the light intensity in a certain place can be interpreted as the probability of finding photons there. On the other hand, particles (electrons, for instance) after passing through a slit distribute themselves in such a way that the probability of finding them in a certain place can be calculated by assuming that they are waves. The notion that there should be any "as if's" in nature is disturbing, and we shall therefore adapt the following point of view:

Matter as well as light consists of particles; however, the behavior of these particles is not described by Newtonian mechanics but by some sort of **wave mechanics** or **quantum mechanics.**

---

[6] Electrons of an energy suitable for this experiment ($\sim 50\ keV$) have according to Eq. 1.5 a very short wavelength. The "replacement" therefore requires a considerable change in the scale of the entire apparatus. Electron diffraction experiments of this kind have become possible only recently and tax severely the skill of even the best equipped experimentalist.

This new kind of mechanics, somehow, does not allow definite predictions for the behavior of an individual particle but describes only some sort of average behavior or, in other words, only the probability that a certain particle will do a certain thing.

The opposite viewpoint is also legitimate: light and matter can be described as waves which, during emission and absorption, manifest themselves as particles.

The theory we are about to develop will encompass both pictures. Which picture we use mentally in thinking of a situation is a matter of practicality or preference—this author, for one, refuses to be considered a wave, no matter how short his wavelength. Obviously the correct quantum mechanical theory must go over into the well-known classical theories of light and matter under certain limiting conditions. The way in which this happens is hinted at by our above experiments. If a vast number of photons arrives during a short time interval, their particle nature will no longer be apparent, but we will observe the intensity predicted by classical electrodynamics. This is especially true if the frequency v and thereby the photon energy hv is small. A single photon of a radio wave contains so little energy that no experimental equipment is sensitive enough to register it. If radio waves are at all detectable, there are so many photons present that the electrodynamical description is fully adequate.

If the mass of a particle is much larger than, let us say, the proton mass, the diffraction effects will become negligible and the probability to find the particle in the zero order diffraction maximum (i.e., straight behind the slit where classical mechanics says it ought to be) will become a certainty.

1.5 HEISENBERG'S UNCERTAINTY RELATION

The thought that for a well-defined initial condition, nature (or at least our theory) should not provide us with a well-defined final state is unpleasant. We shall, therefore, following **Werner Heisenberg,** examine the concept of the well defined initial state more critically.

Obviously we know the initial state of a particle, i.e., its momentum and location, only if we have actually measured it. Let us, therefore, measure in a "*gedanken*-experiment," [7] proposed by **Niels Bohr,** the velocity and location of an electron. Let us assume that we have a microscope so powerful that we can see an individual electron with it. The fundamental process necessary to accomplish this is, of course, that light (i.e., at least one photon) is bounced

---

[7] A *gedanken-* , or thought-experiment, is an imagined experiment which, although impractical or even unfeasible, does not violate any fundamental law of nature. In a *gedanken*-experiment we can let the cow jump over the moon and calculate the initial velocity it needs to do this.

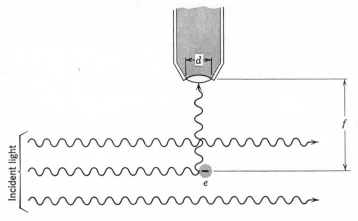

Fig. 1.2   The combined effects of the finite resolution of the microscope and the momentum transferred by the photon set a fundamental limit to the accuracy with which the location and the momentum of the electron can be known *simultaneously*.

off the electron and enters the objective lens of the microscope (see Figure 1.2).

It is well known that, because of the finite size of the diffraction pattern, a microscope of focal length f can resolve a particle position only with an uncertainty:

$$\Delta x \approx f \frac{\lambda}{d} \tag{1.6}$$

where λ is the wavelength of the illuminating light. The photon in bouncing off the electron transfers some momentum to it (Compton effect). The x-component of the momentum of the scattered photon is determined only with an uncertainty

$$\pm p_x \frac{d}{2f}$$

since we do not know where it actually went through the lens. Thus, even if we know the momentum p_x of the photon before the collision perfectly well, after the collision the x-component of its momentum is only known within[8]

$$\Delta p_x = \frac{p_x d}{f} \tag{1.7}$$

---

[8] In the derivation of Eq. 1.7 it has been assumed that the absolute value of the photon momentum has not been changed by the collision, i.e., the electron has been considered to be "heavy" compared with the photon. This assumption is valid as long as the photon energy is small compared with the rest energy, $E_o = 511$ keV, of the electron. For photons of visible light ($E = h\nu \approx 4$ eV) this is certainly a valid assumption.

From the conservation of momentum it follows that after the x-component of the electron-momentum has been observed, it is uncertain within the same limits. The product of the uncertainties of electron x-coordinate and momentum is therefore

$$\Delta x \, \Delta p_x = \left(\frac{f\lambda}{d}\right)\left(\frac{p_x d}{f}\right) = \lambda p_x = \frac{\lambda h}{\lambda} = h \qquad (1.8)$$

(using Eq. 1.4 and assuming that the light was incident in the x-direction, i.e., that $p_v = p_x$).

The fact that

$$\Delta x \, \Delta p_x = h \qquad (1.9)$$

contains neither the particle mass nor the parameters of the microscope suggests that it might be universally valid, as is indeed the case. Equation 1.9 expresses the famous **uncertainty principle** by Heisenberg and says in words that, if the coordinates of any object are known with an accuracy Δx, Δy, Δz, then its momentum is uncertain within[9]:

$$\Delta p_x = \frac{h}{\Delta x}, \qquad \Delta p_y = \frac{h}{\Delta y}, \qquad \Delta p_z = \frac{h}{\Delta z} \qquad (1.10)$$

According to Heisenberg's uncertainty principle as stated in Eq. 1.9, we can never know the *initial state* of any system with complete accuracy, and this rather than a lack of causality is the reason that our theory can determine only the probability that a certain final state will occur.

Much has been written about the question whether nature is really indeterminate or whether the uncertainty principle states merely a limit, although a fundamental one, to the accuracy with which we can measure things. As we have seen in Bohr's *gedanken*-experiment, our trouble stems from the fact that the probe particles (the photons) have a nonzero wavelength and momentum. The uncertainty principle is thus deeply anchored in the wave nature of particles. If a particle with zero wavelength existed, it would not be subject to a description by quantum mechanics, and the uncertainty principle would not apply to it. The particle would also, if used as a probe in Bohr's microscope, destroy the uncertainty principle for all other particles. Until such a particle has been found, we leave the above question to the philosophers.

[9] The alert reader will have noticed that something is amiss here. A single photon does not create an image in the focal plane of the microscope. It would have been better to measure the coordinates of an ensemble of particles using several photons. The result would have been the same. Since a more formal discussion of the uncertainty principle is given later in this book, we shall not pursue this point.

PROBLEMS

1.1 Show that in a collision between a photon and a *free* electron energy *and* momentum would not be conserved if all the photon energy were transferred to the electron (photo effect).

1.2 Calculate the de Broglie wavelength of an electron whose energy is 1000 eV. Compare the result with the wavelength of x-rays with an energy of 1000 eV.

1.3 Calculate the de Broglie wavelength of (a) an electron, (b) a proton having a kinetic energy of 1 eV, 100 eV, 10^6 eV.

1.4 What advantages do electron microscopes have over light microscopes? Why?

1.5 An automobile is moving with a velocity of 60 m/hr. What is its de Broglie wavelength? What, if any, additional assumptions do you have to make to solve this problem?

1.6 Derive an expression for the ratio of electron wavelength/photon wavelength for electrons and photons of equal energy. At what energy are the two wavelengths equal?

1.7 You are given a source of x-rays of 200 keV and are to demonstrate the existence of the Compton effect and the photo effect. What kind of target material, high Z or low Z, will you use as a target in (a) the Compton experiment, and (b) the photoeffect experiment?

1.8 The diameter of an atomic nucleus is of the order of 10^{-13} cm. You want to obtain information about the size and shape of nuclei and you have decided to do this by bombarding them with fast protons. What is the approximate energy to which you have to accelerate the protons?

1.9 A bullet whose mass is 10 g moves with a velocity of 1000 m/sec. In a precision experiment this velocity is determined with an uncertainty of $10^{-4}\%$. How accurately, in principle, could one measure the location of the bullet? Does the uncertainty principle constitute any practical limit on the accuracy with which the location can be determined?

1.10 A high-speed shutter is placed between a monochromatic light source (a laser) and a high resolution spectrometer. First the shutter is held open for a long time while a measurement of the photon energy is made. During a later experiment the shutter is opened for only 10^{-9} sec while the photon energy is measured. How does the result of the second measurement compare with that of the first? (a) Qualitatively, give reasons. (b) Quantitatively, assuming that the spectrometer has infinite resolution.

SOLUTIONS

1.1 Let $E_1 = h\nu$ be the energy of the photon. In this case its momentum will be $p_1 = h\nu/c$. Let $E_2 = mv^2/2 = p_2^2/2m$ be the energy of the electron and p_2 its momentum. Assuming the electron to be at rest initially and assuming that all the photon energy is transferred to the electron, we have

$$h\nu = E_1 = E_2 = \frac{p_2{}^2}{2m}$$

The electron momentum becomes, in this case, $p_2 = \sqrt{2mh\nu} \neq p_1 = h\nu/c$ q.e.d. This is, of course, a non-relativistic calculation; relativistically we obtain a similar *inequality*.

1.8 In a light microscope the wavelength of the illuminating light has to be short compared to dimensions of the object to be observed. Similarly, if we "*illuminate*" a nucleus with a proton, we must require that the de Broglie wavelength of the proton is short compared to the nuclear dimensions if we are to observe any detail. In optical microscopy we settle for nothing less than an image of the object. Nuclear physicists are less demanding and are willing to calculate the size of their object from diffraction patterns. Thus we assume that useful information can be obtained even if the de Broglie wavelength is equal to the nuclear diameter. Hence

$$\lambda = \frac{h}{p} = 10^{-13} \text{ cm}$$

now

$$p = \sqrt{2mE} \quad \text{or} \quad E = \frac{h^2 \cdot 10^{26}}{2m} = \frac{(6.626)^2 \cdot 10^{-54} \cdot 10^{26}}{2 \cdot 1.67 \cdot 10^{-24}} = 13.1 \cdot 10^{-4} \text{ erg}$$
$$= 820 \text{ MeV}$$

This calculation was nonrelativistic. Can you do the same calculation using the proper relativistic connection between energy and momentum? The result should be $E = 617$ MeV.

2

MATTER WAVES

2.1 THE WAVE EQUATION AND THE WAVE FUNCTION

There are several ways to make the transition from classical mechanics to quantum mechanics. They all depend at one point or another on a clever guess of the modifications we have to make in a classical expression to get its quantum-mechanical equivalent.

Our approach to the problem will be to exploit the close resemblance between light waves and particle waves as expressed in Eqs. 1.4 and 1.5 and to guess a plausible-looking wave equation and then to check its validity by applying it to a well-understood experimental situation.

A plane light wave can be described by

$$\varphi(x, t) = E_0 \sin (kx - \omega t) \tag{2.1}$$

where E_0 is the magnitude of the vector of the electric field strength, $\omega = 2\pi\nu$ the angular frequency, and $k = \omega/c = 2\pi/\lambda$ the wave number. Such a wave has the same phase in any plane perpendicular to the direction in which it progresses.

We know from the outcome of the electron diffraction experiment in Chapter 1.3 that a "wave" of electrons of uniform momentum is "monochromatic" (i.e., has a well defined wavelength), so we write in analogy to Eq. 2.1:

$$\psi(x, t) = A \sin (kx - \omega t) \tag{2.2}$$

We do not know whether this is the correct wave function (it will turn out not to be), nor do we know the meaning of the constants A, k, and ω. The interpretation for ω is suggested by Eq. 1.2; we assume that a similar relation exists between the kinetic energy and the frequency of a particle. Hence we try

$$h\nu = \hbar\omega = E = \frac{p_x^2}{2m} \quad \text{or} \quad \omega = \frac{p_x^2}{2m\hbar} \tag{2.3}$$

16

where p_x is the particle momentum.[1] A suitable expression for k comes directly from the experimental result (Eq. 1.5):

$$\lambda = \frac{2\pi}{k} = \frac{h}{p_x} \quad \text{or} \quad k = \frac{p_x}{\hbar} \tag{2.4}$$

Using Eqs. 2.3 and 2.4 in Eq. 2.2 we obtain

$$\psi(x, t) = A \cdot \sin\left(\frac{p_x x}{\hbar} - \frac{p_x^2 t}{2mh}\right) \tag{2.5}$$

This **wave function** is unfortunately not very informative. It describes an infinite plane wave and, hence, does not offer any clue to the location of the particle. Actually this was to be expected. We had specified the particle momentum p_x exactly and should therefore have no idea where the particle is. Nevertheless, we shall try to obtain some hints from Eq. 2.5 concerning the kind of differential equation or **wave function** that describes the motion of particles. The wave function Eq. 2.5 actually satisfies a large number of wave equations, and we try one:

$$\frac{\partial^2 \psi}{\partial t^2} = \alpha \frac{\partial^2 \psi}{\partial x^2} \tag{2.6}$$

This equation is solved by Eq. 2.5. It describes, for instance, the propagation of a plane sound wave through a gas if we identify α with the velocity of sound. Substituting Eq. 2.5 and assuming $\partial p/\partial t = 0$ (i.e., absence of forces acting on the particle), we obtain

$$\frac{-p_x^4}{4m^2\hbar^2} = \frac{-\alpha p_x^2}{\hbar^2} \tag{2.7}$$

or

$$\alpha = \frac{p_x^2}{4m^2} \tag{2.8}$$

This result looks unattractive. We would prefer a wave equation that depends only on basic particle properties such as mass; whereas, dynamic variables such as the momentum should appear only in the wave function.[2] A second look at Eq. 2.7 tells us that p_x would have cancelled if we had differentiated only once with respect to time. So we try a new wave equation:

$$\frac{\partial \psi}{\partial t} = \gamma \frac{\partial^2 \psi}{\partial x^2} \tag{2.9}$$

[1] We have used here $\hbar = h/2\pi \cdot \hbar$ (pronounced $\hbar$-bar) is more frequently used in the modern literature than Plank's original h.
[2] At relativistic velocities, mass becomes a dynamic variable; ours, however, is a non-relativistic theory.

We substitute Eq. 2.5 into Eq. 2.9 and find that this does not end our trouble (Eq. 2.5 is not a solution of Eq. 2.9).

As we can easily verify

$$\psi(x, t) = A \exp\left[i\left(\frac{xp_x}{\hbar} - \frac{p_x^2 t}{2m\hbar}\right)\right] \tag{2.10}$$

is a solution of Eq. 2.9 if

$$\gamma = \frac{i\hbar}{2m} \tag{2.11}$$

Here we stop and review the situation. On the credit side we note:

1. We have found a wave equation (Eq. 2.9) that does not depend explicitly on a dynamic variable.
2. The constant γ contains the fundamental constant $\hbar$ that seems to play such an important role wherever quantum phenomena are concerned.
3. The wave function (Eq. 2.10) which solves Eq. 2.9 describes a "monochromatic" plane wave as was desired.

On the debit side we find:

1. The experimental situation we have tried to describe is rather undemanding. Even though our wave equation can handle a simple plane wave, we still have to test it on a more complicated physical situation.
2. The wave equation (Eq. 2.9) as well as the wave function (Eq. 2.10) are complex. This is particularly disturbing with regard to the latter, since it means that Eq. 2.10 cannot describe a measurable quantity.

So, what does it describe?

We postpone the answer to this question and take another look at Eq. 2.9. Substituting Eq. 2.11 and making the obvious extension to three dimensions we get

$$\frac{\partial \psi}{\partial t} = \frac{i\hbar}{2m} \nabla^2 \psi \quad \text{or} \quad i\hbar \frac{\partial \psi}{\partial t} = \frac{(i\hbar\nabla)^2}{2m} \psi \tag{2.12}$$

If in obtaining Eq. 2.9 we have guessed correctly, Eq. 2.12 will be the wave mechanical equivalent of the classical equation of motion. We compare these two descriptions of the behavior of a free particle:

Classical Equation of Motion   Quantum Mechanical Wave Equation

$$\text{(a)} \ E = p^2 \qquad\qquad \text{(b)} \ i\hbar\frac{\partial \psi}{\partial t} = \frac{(i\hbar\nabla)^2}{2m} \psi \tag{2.13}$$

Written in this form, the two equations exhibit a very formal (and very faint) similarity. This similarity led **Erwin Schrödinger** (1926) to postulate that the transition from the classical description to the quantum mechanical description of a system should be made using the following procedure:

(a) **Write the classical equation of motion in terms of the total energy E, the momentum p and the potential V.**

(b) **Change this equation into an operator equation by replacing E with the operator $ih(\partial/\partial t)$ and by replacing p with the operator[3] $-ih\nabla$**

(c) **Apply the resulting operator equation to a wave function ψ and solve for it.**

This is merely a shrewd guess, and it should be stated that there are other ways to make the transition from classical to quantum mechanics, based on equally tenuous similarities.

The fact that the similarities between Eqs. 2.13a and 2.13b and the conclusions drawn from them are not at all obvious is just another tribute to the genius of Erwin Schrödinger. Before we try to find out about the meaning of the wave function, we shall outline briefly how this theory was completely confirmed by experiment.

The classical equation of motion of a particle in a potential is

$$E = \frac{p^2}{2m} + V(\mathbf{r}) \tag{2.14}$$

Application of the above postulates (b) and (c) yields the **Schrödinger equation**:

$$ih\frac{\partial\psi}{\partial t} = -\frac{\hbar^2}{2m}\nabla^2\psi + V(\mathbf{r})\psi \tag{2.15}$$

In the case of the hydrogen atom,[4] the potential energy is given by Coulomb's law:

$$V(\mathbf{r}) = V(r) = -\frac{e^2}{r} \tag{2.16}$$

substituting this into Eq. 2.15 we obtain

$$ih\frac{\partial\psi}{\partial t} = -\frac{\hbar^2}{2m}\nabla^2\psi - \frac{e^2\psi}{r} \tag{2.17}$$

This is the famous *Schrödinger equation of the hydrogen atom*. We shall solve it later (Chapter 5) and find complete agreement between the experimental values for the energy levels of the hydrogen atom and the values calculated, using Eq. 2.17.

The application of the same procedure to other systems also leads to agreement with the experimental results, and we are today convinced that

[3] This choice of sign will be discussed in Chapter 4.1.
[4] Assuming the proton to be stationary.

Schrödinger's postulates are the key to a complete description of quantum phenomena:

(a) If we can overcome the mathematical difficulties involved in solving the Schrödinger equation.
(b) If we know the force law applicable to the situation.

For more than two particles, the mathematical difficulties are often considerable, just as in the case of the classical many-body problem. There exist, however, very powerful approximative methods to deal with more complicated problems. The exact force law is known to us only for electric and magnetic interactions of the kind existing between a nucleus and its surrounding electrons or between the electrons themselves.[5] The laws that govern nuclear forces are still partly unknown—a fact that further impedes the search for the solution of nuclear many-body problems. We shall see later that quantum mechanics can make qualitative but firm predictions about the outcome of experiments even though we know only the general character of a force (i.e., whether it is attractive, repulsive, spherically symmetric, etc.).

After this sneak preview of events to come, we shall try to understand the role of the wave function in the scheme of things.

Since Eq. 2.10 describes a plane wave filling all space and does not offer any clue, we turn to the Schrödinger equation (Eq. 2.15) for enlightenment. Equation 2.15 is a partial differential equation, and we try to solve it by writing the wave function as a product of a function $u(\mathbf{r})$ that depends only on $\mathbf{r}$ and another function $\varphi(t)$ that depends only on t.

$$\psi(\mathbf{r}, t) = \varphi(t)u(\mathbf{r}) \tag{2.18}$$

hence

$$i\hbar u(\mathbf{r})\frac{d\varphi}{dt} = -\frac{\hbar^2}{2m}\varphi(t)\nabla^2 u(\mathbf{r}) + V(\mathbf{r})\varphi(t)u(\mathbf{r}) \tag{2.19}$$

we divide by $\psi(\mathbf{r}, t)$ and get

$$\frac{i\hbar}{\varphi(t)}\frac{d\varphi}{dt} = -\frac{\hbar^2}{2mu(\mathbf{r})}\nabla^2 u(\mathbf{r}) + V(\mathbf{r}) \tag{2.20}$$

Since the left side of Eq. 2.20 does not depend on $\mathbf{r}$, and the right side does not depend on t, both sides must be equal to the same constant[6], say E.

---

[5] The gravitational force is also well known but is so weak that its manifestations have never been observed on an atomic scale.

[6] The mathematical technique we have used here is called the separation of the variables. It often leads to a simplification of the problem. In our case it allows us to split the partial differential Eq. 2.19 into an ordinary differential equation and a partial differential equation of fewer variables (x, y, and z). A partial differential equation that can be reduced in this manner is said to be separable.

Hence

$$\frac{i\hbar}{\varphi}\frac{d\varphi}{dt} = E \qquad \text{or} \qquad \varphi = \varphi_0 e^{-iEt/\hbar} \tag{2.21}$$

and

$$Eu(\mathbf{r}) = -\frac{\hbar^2}{2m}\nabla^2 u(\mathbf{r}) + V(\mathbf{r})u(\mathbf{r}) \tag{2.22}$$

Equation 2.22 is usually called the **time-independent Schrödinger equation.** It is obvious from Eq. 2.20 that this separation can always be carried out if V is not time dependent, i.e., if $\partial V/\partial t = 0$.

Since $Et/\hbar$ in Eq. 2.21 must be dimensionless, it follows that E has the dimension of an energy; $\omega = E/\hbar$ is a frequency, and since E is a constant, ω must be constant. Thus[7]

$$\psi(\mathbf{r}, t) = \varphi(t)u(\mathbf{r}) = e^{-i\omega t}u(\mathbf{r}) \tag{2.23}$$

This is a wave function that describes a *monochromatic* standing wave whose amplitude u is a function of r.

In Chapter 1, we had interpreted the square of the amplitude of a traveling wave as something proportional to the probability that a photon goes through a unit area in unit time. For a standing wave, the intensity is proportional to the probability of finding a photon in a volume element.

In analogy to this, we interpret[8] $u^*(\mathbf{r})u(\mathbf{r})$ as the **probability density** of finding the particle at $\mathbf{r}$. The square of the absolute value, $u^*(\mathbf{r})u(\mathbf{r})$, was taken to account for the possibility that $u(r)$ might be a complex function.

At this point some clarification of the concept of probability density may be in order.

The probability of finding a point particle at any given point in space is zero because there are infinitely many points in any finite volume. To come to a meaningful definition of the probability of finding a particle somewhere, we have to refer to a finite volume element. The probability of finding a particle in it depends on the distribution of the particles *and* on the size of the volume element.[9] We define as the probability density P the probability w per unit volume:

$$P = \frac{dw}{d\tau} \tag{2.24}$$

to find the particle.

The probability that the particle is in a finite volume V is obviously given

---

[7] The constant φ_0 is now included in the function $u(\mathbf{r})$.
[8] $u^*(\mathbf{r})$ is the complex conjugate of $u(\mathbf{r})$.
[9] The probability of catching a fish depends on how large the net is as well as on the local abundance of fish.

by the integral

$$w = \int_V P \, d\tau \qquad (2.25)$$

taken over this volume. Hence

$$P(\mathbf{r}) = \psi^*(\mathbf{r}, t)\psi(\mathbf{r}, t) = u^*(\mathbf{r})u(\mathbf{r}) \qquad (2.26)$$

Since the particle, if it exists, must be somewhere, the probability density integrated over all volume elements must yield a certainty. Thus

$$\int P(\mathbf{r}) \, d\tau = \int u^*(\mathbf{r})u(\mathbf{r}) \, d\tau = 1 \qquad (2.27)$$

The Schrödinger equation is a homogeneous differential equation and leaves a constant factor in its solutions undetermined. The **normalization of the wave function** Eq. 2.27 allows us to determine this factor and to calculate the absolute probability density.

The probability density can be measured in scattering experiments but, although these experiments confirm the interpretation that we have given the wave function, they are not very precise. The real proof of our theory is rather in the extreme accuracy (6 or more decimal places) with which measured energy values verify its predictions.

Before we continue, we bring our terminology up to date. The time-independent Schrödinger equation (Eq. 2.22)

$$Eu = - \frac{\hbar^2}{2m} \nabla^2 u + Vu \qquad (2.28)$$

is often written as an operator equation (see Appendix A.1):

$$Eu = \left(\frac{-\hbar^2}{2m} \nabla^2 + V \right) u = Hu \qquad (2.29)$$

The operator

$$H = - \frac{\hbar^2}{2m} \nabla^2 + V \qquad (2.30)$$

is called the **Hamilton operator** or the **Hamiltonian** of the problem because of its similarity to the Hamiltonian form of the equation of motion in classical mechanics. Instead of saying that we solve the Schrödinger equation (Eq. 2.28) we often say, "*we find the eigenfunctions u of the Hamiltonian*" (Eq. 2.30) (see Appendix A.1).

The eigenvalues E of the Hamiltonian are the possible energy values of the system. This statement will be made plausible later (Chapter 4.1). In the final analysis, however, it can be justified only through experimental verification, and we add it, belatedly, as **another postulate** to the ones listed on p. 19.

2.2 WAVE PACKETS, MOMENTUM EIGENFUNCTIONS AND THE UNCERTAINTY PRINCIPLE

Having acquired some familiarity with the basic concepts of quantum mechanics, we take another look at the *uncertainty principle*.

It is well known[10] that any nonperiodic function of time $f(t)$ can be expressed as a superposition of sine waves of varying frequency with the help of a **Fourier integral**:

$$f(t) = \frac{l}{\sqrt{2\pi}} \int_{-\infty}^{\infty} A(\omega)e^{i\omega t}\, d\omega \tag{2.31}$$

Similarly a function of r, or for the sake of simplicity x, can be written as

$$\psi(x) = \frac{l}{\sqrt{2\pi}} \int_{-\infty}^{\infty} \varphi(k)e^{ikx}\, dk \tag{2.32}$$

where k is the wave number and $\varphi(k)$ an amplitude depending on it. We recognize the function e^{ikx} as an eigenfunction of the momentum operator since it satisfies the eigenvalue equation

$$-i\hbar \frac{\partial e^{ikx}}{\partial x} = k\hbar e^{ikx} \tag{2.33}$$

The functions e^{ikx} are, therefore, often referred to as momentum eigenfunctions. The eigenvalues $k\hbar$ of e^{ikx} are, by dimension, momenta but whose momenta? To find out, we note that e^{ikx} is not only an eigenfunction of the momentum operator but is also an eigenfunction of the Hamiltonian of a free ($V = 0$) particle.

$$He^{ikx} = \frac{-\hbar^2}{2m} \frac{\partial^2 e^{ikx}}{\partial x^2} = \frac{-\hbar^2 k^2}{2m} e^{ikx} \tag{2.34}$$

According to the postulate on p. 22 the eigenvalues of the Hamiltonian are the possible energy values of the system. We can thus interpret the wave function $\psi(x)$ of a localized particle as a superposition of wave functions of free particles with various momenta. The range of k over which $\varphi(k)$ is substantially different from zero gives the range of momenta $k\hbar$ that we can expect if we make measurements of the particle momentum.

This is very similar to a situation with which we are familiar in another field: electronics. An electric pulse starting at $t = t_1$ and lasting to $t = t_2$, although it is a one-shot event and has no periodicity, can be interpreted as a superposition of sine waves of various frequencies. If such a pulse is put through a circuit whose response is frequency dependent, the circuit will behave exactly as if it had been subjected to a superposition of sine waves covering the frequency range indicated by the Fourier integral.

[10] If not, see Appendixes A.2 and A.3.

Equation 2.32 thus expresses a connection between the spatial eigenfunctions $\psi(x)$ and the momentum eigenfunctions $\varphi(k)$ or, in other words, between the probability that the particle is in a certain place $x \pm \Delta x$ and that it has a certain momentum $p_x \pm \Delta p_x$. Obviously there must, then, exist a connection between Eq. 2.32 and the uncertainty principle. In Appendix A.3 the Fourier integral of a square pulse is derived. It is shown that the spectrum extends to higher and higher frequencies ω as the width of the square pulse is reduced.

If we replace the square pulse $f(t)$ with a square wave $\psi(x)$, we can conclude by comparing Eqs. 2.31 and 2.32 that for a narrow square wave, $\psi(x)$, the "spectrum" of the wave numbers extends to very large values of k.

This would imply that a particle whose wave function is well localized must have a wide momentum spread; however, there is a catch. A square wave is not a solution of the Schrödinger equation. With a little more effort than we have invested in Appendix A.3 one can show, however, that the following theorem holds true.

THEOREM

The smaller the interval $x \pm \Delta x$ is over which a function $\psi(x)$ differs substantially from zero, the larger is the interval k over which its Fourier amplitude $\varphi(k)$ differs substantially from zero.

This theorem whose proof can be found in the literature[11] applies to any kind of function and thereby also to solutions of the Schrödinger equation.

Relying on this theorem, we can now state with confidence that the momentum spectrum of a well localized particle extends to very high momenta. We have thus shown again that the uncertainty principle is deeply rooted in the wave nature of particles.

From the foregoing it is obvious that the momentum spectrum will not only depend on the width but also on the shape of the spatial distribution of the particle. The two curves in Figure 2.1 may illustrate this. They both enclose the same area and have the same full width at half maximum[12] but obviously have different Fourier transforms.

It is interesting to ask for what shape of the spatial distribution, given a width, the width of the momentum distribution is smallest. If we define the uncertainties Δx and Δp_x as the full widths at half maximum of the respective distributions, the answer is: The uncertainty in the momentum is smallest, for a given uncertainty in the location, if the wave function is a gaussian:

$$u(x) = \frac{1}{\sqrt{\sigma \sqrt{\pi}}} e^{-x^2/2\sigma^2} \qquad (2.35)$$

---

[11] For example, L. P. Smith, "*Mathematical Methods for Scientists and Engineers,*" p. 364. Dover Publications Inc., New York, 1961.

[12] A widely used though arbitrary definition of the "width" of a curve.

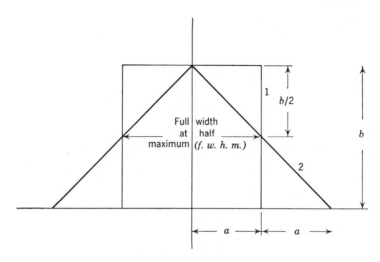

Fig. 2.1 Two curves with the same width ($f.w.h.m$) and the same area can have different Fourier integrals.

The uncertainty product is, in this case;

$$\Delta x \, \Delta p_x = \frac{\hbar}{2} \tag{2.36}$$

This is considerably smaller than the uncertainty product that we had derived from Bohr's *gedanken*-experiment. It is known as the **minimum uncertainty product.** We forego the proof at this point[13] since we shall derive the same result later (Chapter 6.3) in another way.

Thus far we have concerned ourselves only with either monochromatic plane waves or standing waves. The former describe a particle of well-known momentum and completely undetermined location, the latter a particle confined to some region of space—the region where the amplitude $u(\mathbf{r})$ in Eq. 2.22 is different from zero.

As every baseball fan knows, there also exist particles that move in free flight from one place to another, and both their momentum and their position are reasonably well known. We shall now investigate what our theory has to say about this situation. To this end let us assume that the wave function $\psi(x, 0)$ that described the particle at the time $t = 0$ maintains its shape but is shifted along the x-axis as time goes on (see Figure 2.2). A mathematical way of expressing this is to say that

$$\psi(x, t) = \psi(x - vt, 0) \tag{2.37}$$

[13] The proof is found in many of the more advanced texts, e.g., J. Powell and B. Crasemann, *Quantum Mechanics*, Addison and Wesley, Inc., Reading, Mass., 1961, Chapter 3, p. 77.

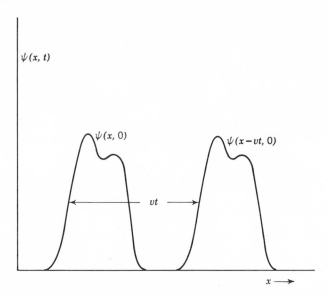

Fig. 2.2 A wave function $\psi(x, t) = \psi(x - vt, 0)$ that maintains its shape while it moves in the positive x-direction. This kind of wave function *cannot* describe a moving particle.

where v is the velocity with which the entire curve is shifted in the positive x direction. This velocity is called the phase velocity.[14]

It is tempting to assume that a moving particle can be described by such a wave function moving with the particle velocity. We shall soon see, however, that this is not true.

The Fourier transform of Eq. 2.37 is obviously

$$\psi(x, t) = \psi(x - vt, 0) = \frac{1}{\sqrt{2\pi}} \int_{-\infty}^{\infty} a(k) e^{ik(x-vt)} \, dk \qquad (2.38)$$

$\psi(x, t)$ is thus decomposed into a packet of plane waves

$$\exp\left[ik(x - vt)\right] = \exp\left[ik\left(x - \frac{k\hbar}{m} t\right)\right] \qquad (2.39)$$

These waves, however, *do not satisfy the Schrödinger equation* (Eq. 2.15). If, on the other hand, we use Schrödinger waves according to Eq. 2.10 to build the wave packet, the resulting wave function will not satisfy Eq. 2.37. We conclude that a wave packet of matter waves cannot propagate in the manner described by Eq. 2.37. Actually this was to be expected. The uncertainty principle tells us that a moving particle whose location we know at

[14] Notice that in this special case the phase velocity would equal the group velocity. For a discussion of the concepts of phase- (wave-) and group velocity, see F. Jenkins and H. White, *Fundamentals of Optics*, Second Edition, McGraw-Hill, New York, 1950.

the time $t = 0$ to be $x \pm \Delta x$, has a momentum distribution covering the range from $p_x - \Delta x$ to $p_x + \Delta x$. Since the momentum is not perfectly defined, the prediction of the future location must contain an additional uncertainty over and above the one the particle had at $t = 0$. In other words, wave packets tend to spread out as time goes by.[15] A quantitative study of this spread in time is given in more advanced texts.[16] It is interesting to compare this situation with the one known in optics. In a light wave moving through vacuum, phase velocity and group velocity are equal. A packet of light waves thus maintains its shape in a vacuum. Only in the presence of a dispersing medium do phase velocity and group velocity differ. The result is known as dispersion. Applying the same terminology to a packet of matter waves which changes its shape constantly, we can say that matter waves show dispersion even in vacuum.

The fact that photons do not undergo dispersion in vacuum does not imply that they are exempt from quantum mechanics. Rather, since they move with the velocity of light regardless of their momentum,

$$p = \frac{\hbar \omega}{c}$$

an uncertainty in momentum does not affect their spread in space.

The group velocity of a packet of light waves is defined as[17]

$$v_g = v_{ph} - \lambda \frac{\partial v_{ph}}{\partial \lambda} \tag{2.40}$$

where v_{ph} is the phase or wave velocity. Equation 2.40 can also be written as[18]

$$v_g = \frac{\partial \omega}{\partial k} \tag{2.41}$$

since for particle waves:

$$\omega = \frac{E}{\hbar} = \frac{p^2}{2m\hbar} = \frac{\hbar^2 k^2}{2m\hbar} \tag{2.42}$$

It follows that

$$v_g = \frac{\partial \omega}{\partial k} = \frac{\hbar k}{m} = \frac{p}{m} = v \tag{2.43}$$

In other words, the group velocity of a packet of matter waves is equal to the velocity of the particle it represents.

[15] This spread may be preceded by a contraction to the minimum size allowed by the uncertainty principle.
[16] For example, J. Powell and B. Crasemann, loc. cit.
[17] See F. Jenkins and H. White, loc. cit.
[18] See Problem 2.7.

PROBLEMS

2.1 Can we measure the wave function of an electron? If so, how? If not, why?

2.2 The normalized wave function of the electron in the lowest energy state of a hydrogen atom is[18]

$$u(r) = Ae^{-me^2r/\hbar^2}$$

(a) Show that this is a solution of the Schrödinger equation (Eq. 2.22) if the potential is a Coulomb potential. (b) Determine the value of the constant A. (c) Find the energy eigenvalue.

2.3 Find a form of the potential for which $u(r) =$ constant is a solution of the Schrödinger equation. Interpret your result.

2.4 An electron whose kinetic energy is 1 eV is trapped in a cubic box of 1 m³ volume with perfectly reflecting walls. What is the probability that the electron can be found in a volume element of 1 cm³ in one of the corners of the box. (*Hint.* Do not calculate, think.)

2.5 Given a homogeneous gravitational force in the $-y$ direction, a frictionless particle moves in a parabolic trough whose cross section is given by $y = x^2$. Write down (a) the time-dependent Schrödinger equation, and (b) the time-independent Schrödinger equation of the system. Assume that there is no motion in the z-direction and that the amplitude is small.

2.6 (a) Normalize
$$u_1(x) = A_1 e^{-\alpha x^2}$$

and

$$u_2(x) = A_2 e^{-\alpha x^2}$$

over the interval $-\infty \leqslant x \leqslant \infty$.
Are these two functions orthogonal over this interval?
(b) Are the functions orthogonal over the interval $0 \leqslant x \leqslant \infty$?

2.7 Derive Eq. 2.41 from Eq. 2.40.

SOLUTIONS

2.6 (a)

$$1 = A_1^2 \int_{-\infty}^{\infty} e^{-2\alpha x^2}\, dx = A_1^2 \sqrt{\frac{\pi}{2\alpha}}$$

hence

$$A_1 = \sqrt{\sqrt{\frac{2\alpha}{\pi}}}$$

$$1 = A_2^2 \int_{-\infty}^{\infty} x^2 e^{-2\alpha x^2}\, dx = \frac{A_2^2}{4\alpha}\sqrt{\sqrt{\frac{2\alpha}{\pi}}}$$

[19] Watch out for the two different meanings of e in this equation.

hence

$$A_2 = \sqrt{4\alpha}\sqrt{\sqrt{\frac{2\alpha}{\pi}}}$$

To find out whether the functions are orthogonal we form

$$\int_{-\infty}^{\infty} u_1(x)u_2(x)\,dx = A_1A_2\int_{-\infty}^{\infty} x\, e^{-2\alpha x^2}\,dx$$

Since the integrand changes sign as we go from x to $-x$ the integral must vanish between symmetrical limits.

2.6 (b) The integral

$$\int_{-\infty}^{\infty} u_1(x)u_2(x)\,dx = A_1A_2\int_{-\infty}^{\infty} xe^{-2\alpha x^2}\,dx = A_1A_2\frac{1}{4\alpha}$$

does not vanish, i.e., the two functions are *not* orthogonal over the interval $0 \leqslant x \leqslant \infty$. It is thus meaningless to say that two functions are orthogonal unless we specify the interval, area, or volume in which they are orthogonal.

3

SOME SIMPLE PROBLEMS

At this point only some mathematics stands between us and our intermediate goal: the calculation of the parameters of the hydrogen atom. A look at the Schrödinger equation in the form in which we shall later solve it (Eq. 5.22) may, however, convince us that we should sharpen our skills on the simpler problems below. These problems do not deal with real physical systems but with simplified abstractions. They will, nevertheless, give some valuable insight into the workings of quantum mechanics.

3.1 THE PARTICLE IN A BOX

For simplicity we consider a one-dimensional box, i.e., we allow a particle to move in a force-free region on the x-axis $-a < x < +a$. At $x = \pm a$, we install a strongly repulsive force in the form of an infinite potential (Figure 3.1).

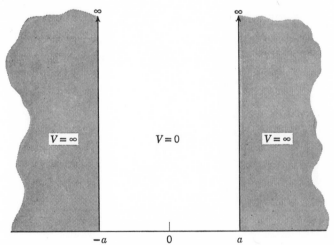

Fig. 3.1 An infinite square well potential.

In the force-free ($V = 0$) interior of the box, the time-independent Schrö-dinger equation (See Eq. 2.28) is

$$Eu = \frac{-\hbar^2}{2m} \cdot \frac{d^2u}{dx^2} \tag{3.1}$$

We try a solution of the form:

$$u(x) = A \sin(\alpha x) + B \cos(\alpha x) \tag{3.2}$$

Substitution into Eq. 3.1 yields

$$E[A \sin(\alpha x) + B \cos(\alpha x)] = \frac{-\hbar^2}{2m}[-A\alpha^2 \sin(\alpha x) - B\alpha^2 \cos(\alpha x)] \tag{3.3}$$

This is an identity[1] if, and only if,

$$\frac{\hbar^2\alpha^2}{2m} = E \quad \text{or} \quad \alpha = \sqrt{\frac{2mE}{\hbar^2}} \tag{3.4}$$

Hence

$$u(x) = A \sin\left(\sqrt{\frac{2mE}{\hbar^2}}\, x\right) + B \cos\left(\sqrt{\frac{2mE}{\hbar^2}}\, x\right) \tag{3.5}$$

is a solution of Eq. 3.1.

This wave function does not tell us very much, since Eq. 3.1 places no restriction on E. Actually this was to be expected, since a particle moving freely along the x-axis can have any energy it pleases. Equation 3.1, however, does not tell the whole story; we have not yet taken into account the influence of the retaining walls.

Outside the region $-a < x < +a$, Eq. 3.1 has to be replaced with

$$Eu(x) = \frac{-\hbar^2}{2m} \frac{d^2u(x)}{dx^2} + Vu(x) \tag{3.6}$$

In the limit $V \to \infty$ this can be satisfied for finite values of E only if

$$u = 0 \quad \text{for all} \quad |x| \geqslant a$$

The infinite potential thus imposes the boundary conditions:

$$u(+a) = u(-a) = 0$$

If we assume the wave function to be continuous at $x = \pm a$, this results in

$$-A \sin(\alpha a) + B \cos(\alpha a) = u(-a) = 0$$

and

$$A \sin(\alpha a) + B \cos(\alpha a) = u(a) = 0 \tag{3.7}$$

---

[1] In other words valid for all values of x.

A trivial solution of Eq. 3.7 is $A = B = 0$, which means that $u(x) \equiv 0$ or that the *box is empty*.

There are, however, two nontrivial solutions; one is

$$A = 0 \qquad \text{hence} \qquad \cos(\alpha a) = 0$$

or

$$u(x) = B \cos(\alpha x) \tag{3.8}$$

where

$$\alpha = \frac{\pi}{2a}, \frac{3\pi}{2a}, \frac{5\pi}{2a}, \dots, \frac{n\pi}{2a}, \text{ etc., and } n \text{ is an odd integer.} \tag{3.9}$$

The other is

$$B = 0 \qquad \text{hence} \qquad \sin(\alpha a) = 0$$

or

$$u(x) = A \sin(\alpha x) \tag{3.10}$$

where

$$\alpha = 0, \frac{\pi}{a}, \frac{2\pi}{a}, \frac{3\pi}{a}, \dots, \frac{n\pi}{a}, \text{ etc., and } n \text{ is an even integer.} \tag{3.11}$$

The *energy eigenvalues* are, according to Eq. 3.4,

$$E = \frac{\alpha^2 \hbar^2}{2m} = \frac{n^2 \pi^2 \hbar^2}{8a^2 m} = E_n; \qquad n = 1, 2, 3, \dots \tag{3.12}$$

i.e., there are infinitely many quadratically spaced energy levels possible.

This leaves only the constants A and B to be determined. If there is one particle in the box, we must have

$$\int_{-\infty}^{\infty} u^2(x)\,dx = \int_{-a}^{a} u^2(x)\,dx = 1 \tag{3.13}$$

Applied to the wave functions (Eqs. 3.8 and 3.10), this **normalization** yields

$$A = B = \frac{1}{\sqrt{a}} \tag{3.14}$$

In Eq. 3.12 we left out the case $n = 0$ for good reasons. According to Eqs. 3.10 and 3.11, $n = 0$ results in $u(x) \equiv 0$. In other words, any particle confined in a box *must* have a certain **minimum energy:**

$$E_1 = \frac{\pi^2 \hbar^2}{8a^2 m} \tag{3.15}$$

This sounds startling, but is a direct *consequence of the uncertainty principle.* If we know the particle coordinates to within $2a$, and we do, then the momentum has to be uncertain to within

$$\Delta p_x = \frac{2\pi \hbar}{2a} \tag{3.16}$$

This amounts to not knowing whether the particle moves with a momentum $|p_x| = \pi\hbar/2a$ in the $+x$ or $-x$ direction. The energy corresponding to this momentum is[2]

$$E = \frac{p_x^2}{2m} = \frac{\pi^2\hbar^2}{8a^2m} = E_1 \tag{3.17}$$

the lowest possible energy state.

This is a simple method to obtain a rough estimate of the minimum energy of a system. It can be used to estimate the energy of electrons in an atom or of nucleons in a nucleus if the atomic or nuclear radius (i.e., the size of the box) is known.

3.2 THE TUNNEL EFFECT

A classical particle (for example, a bowling ball) with kinetic energy E moves toward a potential barrier (position 1, Figure 3.2a). The ball rolls up the slope, and a short time later at $x = 0$ (position 2) all the kinetic energy is transformed into potential energy $V(0) = E$. At this moment the ball reverses its direction and starts to roll back down the slope. Never under any circumstances will it run up the hill any further; in other words, never will it be found in the region $x > 0$, where $V(x) > E$.

Now we investigate what happens under similar circumstances in a quantum mechanical system. We assume that a particle with kinetic energy E approaches a potential barrier of height $V_0 > E$. For mathematical convenience, we assume a discontinuous transition from $V = 0$ to $V = V_0$ at $x = 0$. To the left of the barrier ($x < 0$) the Schrödinger equation reads

$$-\frac{\hbar^2}{2m}\frac{d^2u}{dx^2} = Eu \tag{3.18}$$

and we know the solution already:

$$u = A \sin(\alpha x) + B \cos(\alpha x) \tag{3.19}$$

For $x > 0$ we have

$$\frac{\hbar^2}{2m}\frac{d^2u}{dx^2} = (V_0 - E)u \tag{3.20}$$

The general solution of Eq. 3.20 is

$$u = Ce^{-\gamma x} + De^{\gamma x} \quad \text{with} \quad \gamma = \sqrt{\frac{2m(V_0 - E)}{\hbar^2}} \tag{3.21}$$

[2] The exact agreement is somewhat fortuitous and results from the fact that we have used $\Delta p_x \cdot \Delta x = h$ in the uncertainty relation (Eq. 1.9) instead of, as is more usual, $\Delta x \, \Delta p_x = \hbar$.

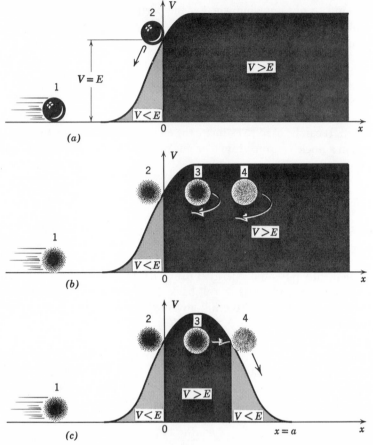

Fig. 3.2 A classical particle (Fig. 3.2*a*) cannot overcome a potential barrier that exceeds its total energy; however, quantum mechanics allows a particle to appear in places that are strictly forbidden to it by classical mechanics (Fig. 3.2*b* and *c*).

In a physically meaningful solution u must be finite for large x.[3] Hence

$$D = 0 \qquad (3.22)$$

At this point we make two additional assumptions: The wave function $u(x)$ *and* its derivative shall be continuous everywhere, including the point $x = 0$.[4] From the continuity at $x = 0$ follows

$$B = C \qquad (3.23)$$

[3] Unless we have a particle source somewhere at $x > 0$ (a possibility that we exclude here). Remember that this solution is valid only for $x > 0$. We do not have to worry about its behavior for negative values of x.

[4] A discontinuity of du/dx or a discontinuity of $u(x)$ would let d^2u/dx^2 go to infinity. According to Eq. 3.20 this can happen only if either E or V becomes infinite.

Thus

$$\frac{du}{dx} = -B\gamma e^{-\gamma x} \qquad \text{for} \qquad x \geqslant 0 \qquad (3.24)$$

$$\frac{du}{dx} = A\alpha \cos(\alpha x) - B\alpha \sin(\alpha x) \qquad \text{for} \qquad x < 0 \qquad (3.25)$$

Hence, from the continuity of du/dx at $x = 0$,

$$-\gamma B = \alpha A \qquad (3.26)$$

The complete solution is thus

$$u(x) = A\left[\sin\left(\sqrt{\frac{2mE}{\hbar^2}}\,x\right) - \sqrt{\frac{E}{V_0 - E}}\cos\left(\sqrt{\frac{2mE}{\hbar^2}}\,x\right)\right] \qquad (3.27)$$

in the region $x < a$, and

$$u(x) = A\sqrt{\frac{E}{V_0 - E}}\exp\left(-\sqrt{\frac{2m(V_0 - E)}{\hbar^2}}\,x\right) \qquad (3.28)$$

in the region $x \geqslant a$. Since $V_0 > E$, (Eq. 3.28) describes a nonperiodic wave whose amplitude decays exponentially.

This means that the particle has a finite—although exponentially de-creasing—probability to be in a region (positions 3 and 4, Figure 3.2b) that is strictly forbidden to it by the laws of classical physics.

If we let the potential go to zero at $x = a$ (Figure 3.2c), particles that have reached position 4 find themselves again in a region with $V < E$ and can continue their journey towards $x = \infty$. This looks as if particles could dig a tunnel through a potential wall of finite thickness and penetrate it if it is too high for them to go over it. This effect is called "tunnel effect" and has important consequences.

As we might suspect, this kind of barrier penetration is not restricted to matter waves. When light is totally reflected at the boundary between glass and air, it actually penetrates into the air. However, in this *forbidden region* the amplitude dies down exponentially over a distance of the order of a wavelength. No light escapes permanently, and we have total reflection. If we bring another piece of glass to within a wavelength of light of the reflecting surface, light can escape into it even though the two surfaces do not touch each other.

It is not difficult to show that other effects familiar from optics occur with matter waves. Light is reflected at both surfaces of a windowpane, i.e., not only in going from a region of low n to one of high n, but also in going from high n to low n. Similarly, matter waves are partially reflected at a potential threshold even if $V_0 < E$ and even in going from a region with $V_0 < E$ to one with $V = 0$. Both these effects are, of course, unknown to classical mechanics. A bowling ball with $E > V_0$ will always roll on toward $x = \infty$, and a bowling ball that approaches the slope in Figure 3.2a from the high side will always roll down.

We shall now discuss an important manifestation of the tunnel effect in the field of nuclear physics. In a nucleus, positively charged particles are held together by attractive nuclear forces of very short range. The exact law obeyed by the nuclear forces is still not completely known. For instance, we do not know whether these forces possess a potential or whether they are velocity dependent. The general situation however, can be described at least approximately with the picture shown in Figure 3.3. Outside a certain distance, R_2, from the center of the nucleus, the long range ($V \propto 1/r$) repulsive Coulomb force between the positively charged particles dominates. Closer in, at R_1, the attractive nuclear force takes over, and we represent it by a potential well. Particles trapped in this well must have a certain minimum kinetic energy, E_1, as a result of the uncertainty principle (see Chapter 3.1). Because of the tunnel effect, an α-particle rattling around in this potential well has a small but finite probability of "tunneling" through the wall. According to our

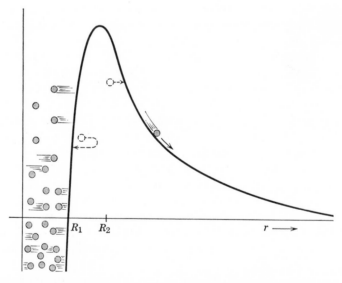

Fig. 3.3 The nucleons in a nucleus are held together in a potential well formed by the nuclear forces. They do not have enough energy to spill over the rim of this well. The tunnel effect allows some of them to get out nevertheless. (Energetically the emission of α-particles is favored over the emission of single nucleons.)

simple-minded theory of this process, we might expect to find an expression like

$$p \propto u^*u = u^2 \propto \exp -2\left[\sqrt{\frac{(\bar{V} - E)2M}{\hbar^2}}(R_1 - R_2)\right] \qquad (3.29)$$

($\bar{V}$ = average potential between R_1 and R_2, M = mass of the α-particle) for the probability of finding the α-particle outside the nucleus (α-decay). A more sophisticated approach to the problem yields

$$p \propto \exp\left(-2\int_{R_1}^{R_2}\sqrt{\frac{2M(V - E)}{\hbar^2}}\, dr\right) \qquad (3.30)$$

which is in satisfactory agreement with experimental results.

Another example of the tunnel effect is the penetration of electrons through a very thin insulating layer between two conductors. This effect is used in certain electronic devices (tunnel diodes).

If the two conductors are superconducting and are held at a slightly different potential, an interesting effect can be observed. In tunneling from the higher potential V_2 through the insulating barrier to the lower potential V_1, an electron looses an amount of potential energy that is given by

$$\Delta E = e(V_2 - V_1) \qquad (3.31)$$

where e is the electron charge. For reasons that are explained by a detailed quantum mechanical theory of superconductivity, electrons can also tunnel through the insulator in pairs. In this case, the excess energy which is now

$$\Delta E = 2e(V_2 - V_1) \qquad (3.32)$$

can be emitted in the form of a photon whose frequency is

$$\omega = \frac{2e(V_2 - V_1)}{\hbar} \qquad (3.33)$$

Not only have such photons recently been observed (the so-called a.c. Josephson effect)[5] but Eq. 3.33 has been found to agree with the experimental results to within 10^{-3} percent.

3.3 THE LINEAR HARMONIC OSCILLATOR

We shall now examine a system that is still a simplified abstraction but that has served as a model for the theoretical study of many real physical situations.

[5] The abbreviation a.c. stands for alternating current. In experiments of this kind, the frequency ω is usually in the microwave range.

It is well known from classical mechanics that a restoring force, which is proportional to the distance from an equilibrium position, $F = -Cx$, leads to a simple harmonic motion. Such a force can be derived from a potential:

$$V = \frac{Cx^2}{2} \tag{3.34}$$

The equation of motion of a one-dimensional classical harmonic oscillator is thus

$$E = \frac{mv^2}{2} + \frac{Cx^2}{2} \tag{3.35}$$

We mention an obvious way to realize such a system: A mass, M, held between two springs obeying Hooke's law (see Figure 3.4). A solution of the equation of motion Eq. 3.35 is

$$x = A \sin(\omega t) \tag{3.36}$$

We determine the constants:

$$E = \frac{mA^2}{2} \omega^2 \cos^2(\omega t) + \frac{CA^2}{2} \sin^2(\omega t) \tag{3.37}$$

or

$$\frac{2E}{A^2C} = \frac{m\omega^2}{C^2} \cos^2(\omega t) + \sin^2(\omega t) \tag{3.38}$$

This identity holds true only if

$$\omega^2 = \frac{C}{m} \quad \text{and} \quad A^2 = \frac{2E}{C} \tag{3.39}$$

hence

$$x(t) = \sqrt{\frac{2E}{C}} \sin\left(\sqrt{\frac{C}{m}} t\right) \tag{3.40}$$

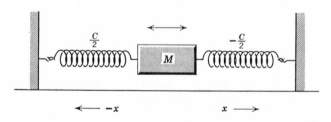

Fig. 3.4 In a classical harmonic oscillator a particle is bound to an equilibrium position by a force that is proportional to the distance from this position. Another way to express the *same fact* is to say that the potential energy is proportional to the square of the distance from the equilibrium position.

In other words, a classical harmonic oscillator in one dimension can have any amplitude and, hence, any total energy but only one frequency.

$$\omega = \sqrt{\frac{C}{m}} \tag{3.41}$$

Now we write the Schrödinger equation for the one-dimensional quantum mechanical harmonic ascillator. (We shall show later that the three-dimensional Schrödinger equation can be separated into three one-dimensional equations.)

$$Eu = -\frac{\hbar^2}{2m}\frac{d^2u}{dx^2} + \frac{\omega^2 m x^2}{2}u \tag{3.42}$$

We have used here $\omega^2 = C/m$, i.e., we have expressed the potential energy in terms of the frequency ω that a classical harmonic oscillator with the same restoring force C would have. To simplify Eq. 3.38, we rewrite

$$\frac{2Eu}{\hbar\omega} = \lambda u = -\frac{\hbar}{m\omega}\frac{d^2u}{dx^2} + \frac{m\omega x^2}{\hbar}u \tag{3.43}$$

This simplifies if we measure x in units $\sqrt{\hbar/m\omega}$,[6] i.e., substitute $x = x'\sqrt{\hbar/m\omega}$:

$$\lambda u = -\frac{d^2u}{dx'^2} + x'^2 u \tag{3.44}$$

For convenience, we shall drop the primes henceforth and write x instead, keeping in mind that our unit of length is now: $\sqrt{\hbar/m\omega}$. Innocent though it looks, Eq. 3.44 is not easy to solve. We notice, however, that for very large values of x, $\lambda u \ll x^2 u$, so that the Schrödinger equation becomes

$$\frac{d^2u_\infty}{dx^2} = x^2 u_\infty \tag{3.45}$$

For large x this equation has the solution:

$$u_\infty \approx x^n e^{-x^2/2} \tag{3.46}$$

This is usually called the **asymptotic solution** of Eq. 3.44. We split a factor $e^{-x^2/2}$ off the eigenfunction $u(x)$ and substitute

$$u(x) = e^{-x^2/2}f(x) \tag{3.47}$$

into Eq. 3.44, hoping that this will simplify our problem.[7] We form

$$\frac{d^2u}{dx^2} = e^{-x^2/2}[-f(x) - 2xf'(x) + f''(x) + x^2 f(x)] \tag{3.48}$$

---

[6] We can easily convince ourselves that this has, indeed, the dimension of a length.
[7] It will, see Problem 3.4.

Substituting Eqs. 3.47 and 3.48 into Eq. 3.44, we obtain

$$\lambda f(x) = f(x) + 2xf'(x) - f''(x) - x^2f(x) + x^2f(x) \tag{3.49}$$

or

$$f''(x) - 2xf'(x) + f(x)(\lambda - 1) = 0 \tag{3.50}$$

In order to solve this new differential equation, we expand

$$f(x) = \sum_0^\infty A_k x^k \tag{3.51}$$

hence

$$xf'(x) = \sum_1^\infty k A_k x^k \tag{3.52}$$

$$f''(x) = \sum_2^\infty k(k-1)A_k x^{k-2} = \sum_0^\infty (k+2)(k+1)A_{k+2} x^k \tag{3.53}$$

We substitute Eq. 3.51, 3.52, and 3.53 into Eq. 3.50.

$$\sum_0^\infty (k+2)(k+1)A_{k+2} x^k - 2\sum_0^\infty k A_k x^k + (\lambda - 1)\sum_0^\infty A_k x^k \equiv 0 \tag{3.54}$$

This is an identity only if

$$(k+2)(k+1)A_{k+2} - 2A_k k + (\lambda - 1)A_k = 0 \tag{3.55}$$

that is, if the coefficients of x^k vanish separately for all values of k. Equation 3.55, therefore, gives the following recurrence relation[8] for the A_k:

$$A_{k+2} = \frac{2k + 1 - \lambda}{(k+2)(k+1)} A_k \tag{3.56}$$

This means that any series whose coefficients satisfy Eq. 3.56 will satisfy Eq. 3.50. As usual, this includes many more solutions than the physical meaning of Eq. 3.50 calls for and, as usual, we invoke the boundary conditions to sift out the physically meaningful solutions.

In the linear harmonic oscillator the particle can go to infinity if its energy E is infinite. The oscillator has thus no clearly defined boundary, and the only condition that we can impose in good conscience is

$$\int_{-\infty}^{\infty} u^*(x)u(x)\, dx = 1 \tag{3.57}$$

This implies that $u^*(x)u(x)$ goes to zero faster than $|1/x|$ since the latter condition would still lead to a logarithmic singularity. With this in mind, we examine Eq. 3.56.

$$\lim_{k \to \infty} \frac{A_{k+2}}{A_k} = \frac{2}{k} \tag{3.58}$$

[8] See Problem 3.7.

A series with coefficients that satisfy Eq. 3.58 is convergent,[9] but convergence alone is not enough. Unless

$$f(x) = \sum_0^\infty A_k x^k \tag{3.59}$$

stays smaller than $e^{+x^2/2}$ for large values of x,

$$u(x) = e^{-x^2/2} f(x) \tag{3.60}$$

will at best approach a constant, and Eq. 3.57 cannot be satisfied. However, as we shall see presently $f(x)$ in Eq. 3.59 *does not* stay smaller than $e^{x^2/2}$. We expand

$$e^{x^2/2} = \sum_0^\infty \frac{x^{2k}}{2^k k!} = \sum_{k\text{-even}} \frac{x^k}{2^{k/2}(k/2)!} \tag{3.61}$$

The ratio of two consecutive coefficients becomes for large values of k:

$$\frac{2^{k/2}(k/2)!}{2^{k/2+1}(k/2+1)!} = \frac{1}{2(k/2+1)} \approx \frac{1}{k}$$

Since the same ratio for Eq. 3.58 approaches $2/k$, it follows that $f(x)$ in Eq. 3.59 *does not* stay smaller than $e^{x^2/2}$ for large x. Fortunately, there is a way out of this dilemma. If in Eq. 3.56 for a certain integer $k = n$,

$$2n + 1 = \lambda \tag{3.62}$$

A_{n+2} and hence all following coefficients vanish. In this case we get

$$f(x) = \sum_0^n A_k x^k \tag{3.63}$$

This is a polynomial and therefore certainly smaller than $e^{x^2/2}$ for large values of x and finite n. The normalization condition (Eq 3.57) can therefore be satisfied only if

$$\lambda = 2n + 1 \tag{3.64}$$

and if, at the same time (depending on whether n is even or odd), either all the odd-numbered or all the even-numbered A_k vanish. We thus have two series of eigenfunctions with ascending values of the **quantum number** n:

$$u_n(x) = e^{-x^2/2} \sum_0^n A_k x^k \qquad \begin{array}{l} n = 0, 2, 4, \ldots \\ \\ n = 1, 3, 5, \ldots \end{array} \tag{3.65}$$

one with even, and one with odd values of n. The A_k are given by the recurrence relation Eq. 3.56 *if* we know A_0 and A_1. We shall now derive some eigenfunctions explicitly. To this end, we assume that all odd-numbered

---

[9] This is the well-known ratio test. See for instance, V. Kaplan, *Advanced Calculus*, Addison-Wesley Inc., Reading, Mass., 1953.

$A_k = 0$ and that $n = 0$ (i.e., that our series breaks off after the first term). From Eq. 3.56 follows $\lambda = 1$; hence

$$A_0 = A_{00}, \quad \text{and} \quad A_{02} = 0 = A_{04} = A_{06}, \text{ etc.}$$

We write the constants A with two subscripts to account for the possibility that they might have different values for a series that breaks off after n terms then they have for a series that breaks off after n' terms. Now we go to $n = 2$, $\lambda = 5$, and get

$$A_{20}, A_{22} = -2A_{20} \quad 0 = A_{24} = A_{26} = \text{etc.},$$

$$n = 4 \quad \lambda = 9$$

$$A_{40}, A_{42} = -4A_{40}, A_{44} = \tfrac{4}{3}A_{40} \quad 0 = A_{40} = A_{48}, \text{ etc.}$$

Next we derive some odd eigenfunctions:

$$n = 1 \quad \lambda = 3$$

$$A_{11}, \quad A_{13} = 0, \text{ etc.}$$

$$n = 3 \quad \lambda = 7$$

$$A_{31}, \quad A_{33} = -\tfrac{2}{3}A_{31} \quad 0 = A_{35} = A_{37} = \text{etc.}$$

$$n = 5 \quad \lambda = 11$$

$$A_{51}, A_{53} = -\tfrac{4}{3}A_{51}, A_{55} = \tfrac{4}{15}A_{51} \quad 0 = A_{57} = \text{etc.}$$

The polynomials whose coefficients we have just derived are called Hermite polynomials. The Hermite polynomials have been well investigated, and we list here, without proof, some convenient formulas to determine their coefficients:

$$H_n(x) = (-1)^n e^{-x^2} \frac{d^n}{dx^n}\left(e^{-x^2}\right) \tag{3.66}$$

$$H_n(x) = 2^n x^n - 2^{n-1}\binom{n}{2}x^{n-2} + 2^{n-2} \cdot 1 \cdot 3\binom{n}{4}x^{n-4}$$

$$-2^{n-3} \cdot 1 \cdot 3 \cdot 5\binom{n}{6}x^{n-6} + \cdots \tag{3.67}$$

$$\frac{dH_n(x)}{dx} = 2nH_{n-1}(x) \tag{3.68}$$

$$H_{n+1}(x) = 2xH_n(x) - 2nH_{n-1}(x) \tag{3.69}$$

In keeping with convention, we have used in Table 3.1, which lists the first six eigenfunctions, the coefficients as they are given by the above formulas. We notice that the ratio of the coefficients is the same as that given by Eq. 3.56. The normalization, however, will be changed and we shall now call them

Table 3.1

n	λ_n	E_n	u_n
0	1	$\dfrac{\hbar\omega}{2}$	$B_0 e^{-x^2/2}$
1	3	$\dfrac{3\hbar\omega}{2}$	$B_1 2x e^{-x^2/2}$
2	5	$\dfrac{5\hbar\omega}{2}$	$B_2(4x^2 - 2)e^{-x^2/2}$
3	7	$\dfrac{7\hbar\omega}{2}$	$B_3(8x^3 - 12x)e^{-x^2/2}$
4	9	$\dfrac{9\hbar\omega}{2}$	$B_4(16x^4 - 48x^2 + 12)e^{-x^2/2}$
5	11	$\dfrac{11\hbar\omega}{2}$	$B_5(32x^5 - 160x^3 + 120x)e^{-x^2/2}$

B_n. Also, again in keeping with convention, we have chosen the sign of the Hermite polynomials in Table 3.1 and Figure 3.5 so that the highest power of x is always positive. This amounts to a choice of the phase of the wave function which has no physical significance.

The constants B_n can be obtained from the normalization.

Example

$$B_3{}^2 \int_{-\infty}^{\infty} (8x^3 - 12x)^2 e^{-x^2} \, dx = 1 \tag{3.70}$$

The normalization as defined in the example (Eq. 3.70) yields, in the general case,[10]

$$B_n = \frac{1}{\sqrt{2^n n!} \sqrt{\pi}} \tag{3.71}$$

[10] For a derivation of this expression, see J. L. Powell and B. Crasemann, *Quantum Mechanics*, Addison-Wesley, Inc., Reading, Mass. 1961.

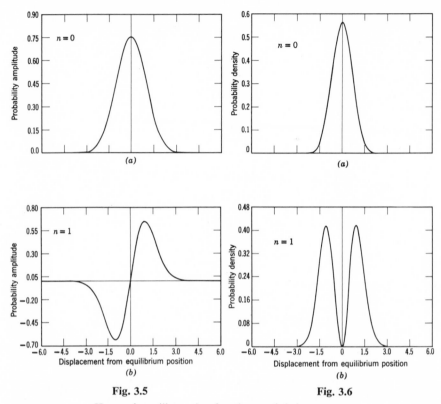

Fig. 3.5   **Fig. 3.6**

Harmonic oscillator eigenfunctions and their squares.

Figures 3.5 and 3.6 show the normalized wave functions and their squares for the quantum numbers, $n = 0$ to $n = 7$.

We sum up: The boundary conditions select, just as in the case of the particle in the box, a denumerable set of eigenfunctions belonging to a set of discrete, (in this case, equidistant) energy eigenvalues.

The eigenfunctions form two discrete sets; one remains unchanged under a mirror transformation (i.e., if we change x to $-x$), and the other changes sign.

Functions with this kind of behavior under exchange of x with $-x$ are said to have a definite **parity.** If

$$u(\mathbf{r}) = u(-\mathbf{r}) \tag{3.72}$$

the parity is said to be even, and if

$$u(\mathbf{r}) = -u(-\mathbf{r}) \tag{3.73}$$

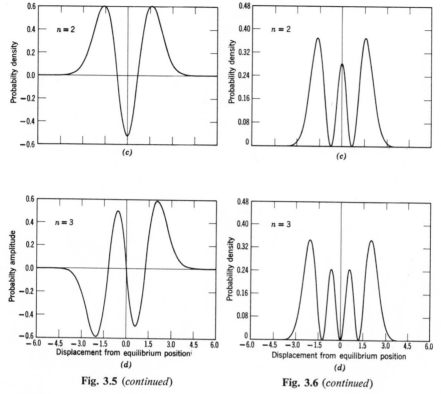

Fig. 3.5 *(continued)*   Fig. 3.6 *(continued)*

the parity is said to be odd.

Examples

$$\cos x,\ x^2,\ x^2 + x^4 + a$$

have even parity;

$$\sin x,\ x,\ x^3$$

have odd parity.

$$e^x,\ x + x^2$$

do not have a definite parity. We shall learn more about the significance of the parity of an eigenfunction in subsequent chapters (see Chapter 6.6).

It is now easy to solve the harmonic oscillator problem in three dimensions. The Hamiltonian is

$$H = -\frac{\hbar^2}{2m}\nabla^2 + \frac{kr^2}{2} \tag{3.74}$$

measuring, as before, lengths in units

$$\sqrt{\frac{\hbar}{m\omega}}$$

and using

$$E = \frac{\hbar\lambda\omega}{2} \tag{3.75}$$

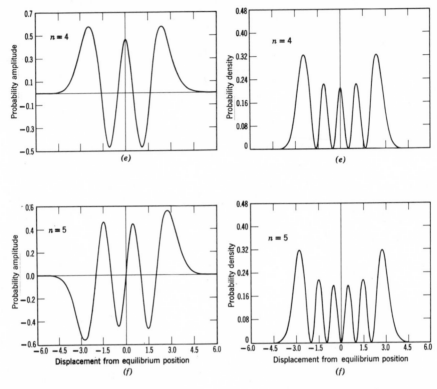

Fig. 3.5 (*continued*)   Fig. 3.6 (*continued*)

we obtain the Schrödinger equation:

$$\nabla^2\psi + (\lambda - r^2)\psi = 0 \qquad (3.76)$$

We try a solution of the form:

$$\psi(x, y, z) = u(x)v(y)w(z) \qquad (3.77)$$

$$wv\frac{\partial^2 u}{\partial x^2} + uw\frac{\partial^2 v}{\partial y^2} + uv\frac{\partial^2 w}{\partial z^2} + (\lambda - x^2 - y^2 - z^2)uvw = 0 \qquad (3.78)$$

We divide by uvw

$$\frac{1}{u}\frac{\partial^2 u}{\partial x^2} + \frac{1}{v}\frac{\partial^2 v}{\partial y^2} + \frac{1}{w}\frac{\partial^2 w}{\partial z^2} + (\lambda - x^2 - y^2 - z^2) = 0 \qquad (3.79)$$

Obviously, this can be written so that either all the x-dependent terms, or all the y-dependent terms, or all the z-dependent terms are on one side.

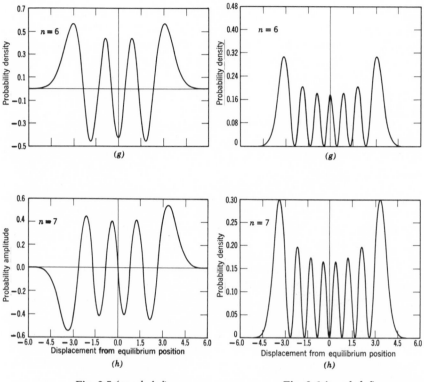

Fig. 3.5 (*concluded*)   Fig. 3.6 (*concluded*)

Hence, the equation can be separated by using three different separation constants, λ_x, λ_y, λ_z, which have to satisfy

$$\lambda_x + \lambda_y + \lambda_z = \lambda \tag{3.80}$$

This gives us three independent equations:

$$\frac{d^2u}{dx^2} + (\lambda_x - x^2)u(x) = 0$$

$$\frac{d^2v}{dy^2} + (\lambda_y - y^2)v(y) = 0 \tag{3.81}$$

$$\frac{d^2w}{dz^2} + (\lambda_z - z^2)w(z) = 0$$

The solutions for these one-dimensional equations have already been obtained and are listed in Table 3.1.

To sum it up, the wave function of the three-dimensional harmonic oscillator is the product of the wave functions of three one-dimensional harmonic oscillators in the x, y, and z direction. The total energy is the sum of the energies of the three one-dimensional oscillators. This results in a very complicated behavior since a solution for the x-coordinate with, for example, $n = 2$ can combine with any solution for the y and z coordinate.

It is interesting to compare the probability of finding the particle in a certain location in the potential well for the classical and quantum-mechanical harmonic oscillator.

For the classical harmonic oscillator the probability is, obviously, inversely proportional to the velocity that the particle has when it moves through a certain volume element (length element in the one-dimensional case). Hence

$$ P \, dx \propto \frac{dx}{v} \propto \frac{dx}{\sqrt{A^2 - x^2}} \tag{3.82} $$

For the quantum-mechanical harmonic oscillator, it is

$$ P \, dx \propto u^2(x) \, dx \tag{3.83} $$

Figure 3.7 compares the probability density of the particle at various distances from the origin in both a quantum-mechanical harmonic oscillator of quantum number $n = 60$ and a classical harmonic oscillator of the same total energy.

There is an excellent agreement between the classical probability density and the average probability density in the quantum-mechanical case. Comparing Figure 3.6 with Figure 3.7, we see that this agreement develops gradually as n increases. This is a manifestation of the **correspondence principle:** Quantum-mechanical systems, in general, approach the classical behavior as we go to large values of the quantum numbers.

Figure 3.8 shows the energy levels and the probability densities of the first eleven states in relation to the harmonic oscillator potential: $V = Cx^2/2$. The quantum-mechanical harmonic oscillator is a reasonable approximation of many physical systems. As an example, we mention diatomic molecules. In a diatomic molecule the two atoms can vibrate around their equilibrium position. For vibrations of small amplitude, the restoring force is very nearly proportional to the amplitude and the molecule resembles a harmonic oscillator rather closely. Consequently the vibrational states of a diatomic molecule lead to nearly equidistant spectral lines.

Historically the harmonic oscillator has played an important role in the development of quantum mechanics. Guided by experimental evidence from the blackbody radiation, Max Planck postulated in 1900 that a harmonic oscillator should only be able to absorb or emit radiation in discrete amounts. This was the beginning of the era of the "old quantum mechanics."

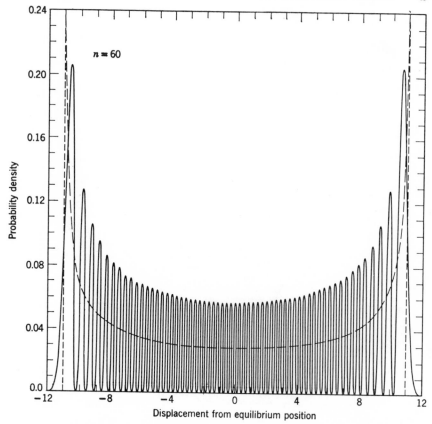

Fig. 3.7 For large values of n the average value of the probability density closely resembles the classical probability density (dotted line).

In 1925, Heisenberg invented a new kind of "matrix mechanics" and showed that, applied to the harmonic oscillator, it, indeed, led to discrete equidistant energy levels.

3.4 THE ONE-DIMENSIONAL CRYSTAL[11]

A metal crystal consisting of a vast number of positive ions and electrons is certainly a quantum-mechanical many-body system par excellence. Fortunately it is possible to derive some of its salient features from the following drastically simplified model.

---

[11] The material in this section is somewhat more difficult than that in the preceding sections. Chapter 3.4 is selfcontained, however, and its study can be postponed until later.

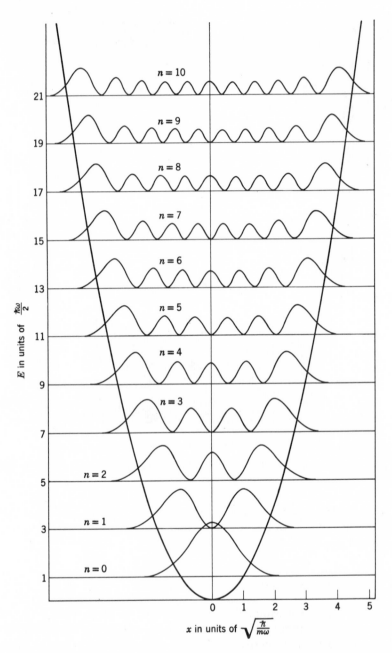

Fig. 3.8 The quantum-mechanical probability density of finding a harmonically bound particle at various distances from the equilibrium position. Notice that, especially for small values of n, there is a substantial probability of finding the particle outside the region allowed to it by classical mechanics, i.e., the parabola.

(a) The *positive ions* are much heavier than the electrons and are known, experimentally, to be almost immovable. We assume that they are stationary and that they *provide a periodic potential* with a period d in which the electrons move.

(b) The *electrons* are very nimble, and able to move out of each others way. We assume that the electric *potential* created by the electrons is the *same everywhere* in the crystal.

These assumptions reduce our problem to the study of the motion of an individual electron in a superposition of a constant and a periodic potential. To further ease the solution, we make some *mathematical simplifications*.

(c) We consider only one dimension, i.e., we investigate the motion of an electron in the one-dimensional crystal lattice of Figure 3.9.

(d) Since the absolute value of the potential has no influence on the motion of the electron, we assume the potential to be zero halfway between lattice points.

(e) To avoid trouble that might arise at the two ends of the crystal, we join them to form a ring.

Before we set out to solve the Schrödinger equation for this problem, we try to find out as much as we can about the general character of its solutions. The crystal potential was assumed to be periodic with a period d:

$$V(x + d) = V(x) \qquad (3.84)$$

The fact that the shape of the potential is the same in all cells of our one-dimensional crystal lattice does not mean that the eigenfunctions have to be

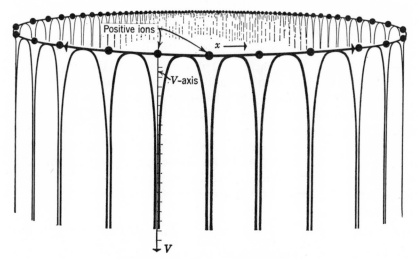

Fig. 3.9 A one-dimensional crystal ring.

periodic with the lattice period. True, the Schrödinger equation remains unchanged if we replace $V(x)$ with $V(x + d)$, but it leaves a constant factor in its eigenfunctions undetermined. At the point $x + d$, the Schrödinger equation is, therefore, the same as it was at the point x. Its solution, $u(x + d)$, could, however, just as well be

$$u(x + d) = \mu_1 u(x) \tag{3.85}$$

where μ_1 is a (possibly complex) constant. If we move over one more lattice space, the same argument holds true. Hence

$$u(x + 2d) = \mu_1 \cdot \mu_2 u(x) \tag{3.86}$$

If we continue this N times (where N is the number of ions in our crystal ring), we get back to where we came from. Hence

$$u(x + Nd) = \mu_1 \cdot \mu_2 \cdots \mu_N u(x) = u(x) \tag{3.87}$$

or

$$\mu_1 \cdot \mu_2 \cdots \mu_N = 1 \tag{3.88}$$

Because of the symmetry of the crystal ring, it cannot make any difference at which lattice point we started, and we conclude that all the μ_k must be equal. Hence

$$u(x + Nd) = \mu^N u(x) = u(x) \tag{3.89}$$

or

$$\mu = e^{2\pi i n/N}, \qquad n = 0, 1, 2, \ldots, N - 1 \tag{3.90}$$

We have, therefore,

$$u(x + md) = e^{2\pi i n m/N} \cdot u(x) \tag{3.91}$$

The statement made in Eq. 3.91 can also be expressed in the following form:

$$u(x) = e^{2\pi i n x/Nd} \cdot w(x) = e^{ikx} \cdot w(x) \tag{3.92}$$

where $k = 2\pi n/Nd$ and where $w(x)$ is periodic with the period d [i.e., $w(x + d) = w(x)$]. The proof is easily accomplished by replacing x with $x + md$ in Eq. 3.92. Equation 3.92 is known to physicists as Bloch's theorem[12] and to mathematicians as Floquet's theorem.[13] It can easily be extended to three dimensions.

The plane wave described by Eq. 3.92 has an amplitude $w(x)$ that is periodic with the period d of the lattice. As we had emphasized earlier, $u(x)$ itself need not have the same periodicity. However, it is easy to convince ourselves that *the probability density, $u^*(x)u(x)$, will be periodic with the lattice period.* This follows from

$$u^*(x + d)u(x + d) = w^*(x + d)w(x + d) = w^*(x)w(x) = u^*(x)u(x) \tag{3.93}$$

[12] Felix Bloch, *Z.f. Physik*, **52**, 555 (1928).
[13] A. M. G. Floquet, *Ann. del' Ecole norm. sup.*, **2**, XII (1883).

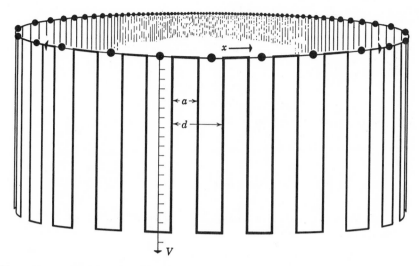

Fig. 3.10   For reasons of mathematical convenience the ring of Fig. 3.9 is idealized by a succession of square well potentials.

If we want to go on with the solution of the Schrödinger equation, we have to specify the actual shape of the periodic potential. To simplify things, we assume the physically impossible though mathematically convenient Kronig Penney potential[14] (shown in Figure 3.10). We have shown that the solutions of the Schrödinger equation with a periodic potential:

$$\frac{d^2u}{dx^2} + \frac{2m}{\hbar^2}(E - V)u = 0 \tag{3.94}$$

are of the form:

$$u(x) = e^{ikx}w(x) \tag{3.95}$$

Substitution of Eq. 3.95 into Eq. 3.94 yields a differential equation for $w(x)$:

$$\frac{d^2w}{dx^2} + 2ik\frac{dw}{dx} + \frac{2m}{\hbar^2}\left(E - \frac{k^2\hbar^2}{2m} - V\right)w = 0 \tag{3.96}$$

To solve Eq. 3.96 we substitute

$$w(x) = e^{i\gamma x} \tag{3.97}$$

The resulting identity yields[15]

$$\gamma = -k \pm \sqrt{\frac{2m(E - V)}{\hbar^2}} \tag{3.98}$$

[14] R. de L. Kronig and W. G. Penney, *Proc. Roy. Soc.*, **A130**, 499 (1931).
[15] Remember that in the Kronig Penney model, V is (except for the discontinuities) a constant.

In the region of zero potential, $0 < x < a$, the general solution of Eq. 3.96 becomes, therefore,

$$w_0(x) = Ae^{i(\alpha - k)x} + Be^{-i(\alpha - k)x} \tag{3.99}$$

where

$$\alpha = +\sqrt{\frac{2mE}{\hbar^2}} \tag{3.100}$$

In the potential well, $a < x < d$, the general solution is

$$w_v(x) = Ce^{i(\beta - k)x} + De^{-i(\beta - k)x} \tag{3.101}$$

where

$$\beta = \sqrt{\frac{2m(E - V)}{\hbar^2}} \tag{3.102}$$

If we require that the eigenfunction $u(x)$ and its first derivative be continuous[16] at $x = 0$ and $x = a$, we get

$$w_0(0) = w_v(0), \qquad w_0(a) = w_v(a) \tag{3.103}$$

$$\left(\frac{dw_0}{dx}\right)_0 = \left(\frac{dw_v}{dx}\right)_0, \qquad \left(\frac{dw_0}{dx}\right)_a = \left(\frac{dw_v}{dx}\right)_a \tag{3.104}$$

These four *linear homogeneous equations* are sufficient to determine the four constants A, B, C, and D *if* the determinant of their coefficients vanishes. This fourth order determinant of transcendental functions of E is difficult to solve for E, and we simplify the problem even further. We shrink the region $a < x < d$, increasing at the same time V so that the product:

$$V(d - a) = K \tag{3.105}$$

remains constant. Since $u(x)$ was to be continuous at $x = a$ and $x = d$, we can integrate Eq. 3.96 if we narrow the potential well so much that $u(x)$ remains (practically) constant in going from $x = a$ to $x = d$. With the potential well this narrow the condition (3.105) makes V so large that we can neglect both k and E inside the well. Hence Eq. 3.96 becomes

$$\frac{d^2w}{dx^2} + 2ik\frac{dw}{dx} - \frac{2mVw}{\hbar^2} = 0 \tag{3.106}$$

We integrate Eq. 3.106 over the region of the potential well:

$$\int_a^d \frac{d^2w}{dx^2}\,dx + 2ik\int_a^d \frac{dw}{dx}\,dx - \frac{2m}{2\hbar}\int_a^d Vw\,dx = 0 \tag{3.107}$$

this yields

$$\left(\frac{dw}{dx}\right)_d - \left(\frac{dw}{dx}\right)_a + 2ik\big(w(d) - w(a)\big) - \frac{2m}{\hbar^2}w(d)K = 0 \tag{3.108}$$

[16] See footnote 4.

Because of the periodicity of $w(x)$ and its continuity at $x = 0$

$$w(d) = w(0) = w(a) \tag{3.109}$$

so that the second integral vanishes. Hence

$$\left(\frac{dw}{dx}\right)_0 - \left(\frac{dw}{dx}\right)_a - \frac{2m}{\hbar^2} w(0)K = 0 \tag{3.110}$$

$$\left(\frac{dw}{dx}\right)_0 \quad \text{and} \quad \left(\frac{dw}{dx}\right)_a$$

are the derivatives on either side of the potential well, i.e., in the region of zero potential, therefore

$$\left(\frac{dw_0}{dx}\right)_0 - \left(\frac{dw_0}{dx}\right)_a - \frac{2m}{\hbar^2} w_0(0)K = 0 \tag{3.111}$$

Together with Eq. 3.99 this yields

$$Ai(\alpha - k)(1 - e^{i(\alpha-k)a}) - Bi(\alpha + k)(1 - e^{-i(\alpha+k)a}) - \frac{2m}{\hbar^2} K(A + B) = 0 \tag{3.112}$$

Equations 3.99 and 3.109 yield

$$A + B = Ae^{i(\alpha-k)a} + Be^{-i(\alpha+k)a} \tag{3.113}$$

Equations 3.112 and 3.113 are two linear homogeneous equations for A and B. They can be solved if the determinant of the coefficients vanishes:

$$\begin{vmatrix} 1 - e^{i(\alpha-k)a} & 1 - e^{-i(\alpha+k)a} \\ i(\alpha - k)(1 - e^{i(\alpha-k)a}) - \dfrac{2mK}{\hbar^2} & -i(\alpha + k)(1 - e^{-i(\alpha+k)a}) - \dfrac{2mK}{\hbar^2} \end{vmatrix} = 0 \tag{3.114}$$

Multiplying this out and using the identity $e^{i\varphi} = \cos \varphi + i \sin \varphi$, we get a transcendental equation for α:

$$\frac{Km}{\lambda \hbar^2} \sin (\alpha a) + \cos (\alpha a) = \cos (ka) \tag{3.115}$$

or substituting the value of α given Eq. 3.100 and replacing[17] a with the lattice constant d:

$$\frac{Km}{\hbar\sqrt{2mE}} \sin \left(\frac{d}{\hbar} \sqrt{2mE}\right) + \cos \left(\frac{d}{\hbar} \sqrt{2mE}\right) = \cos (kd) \tag{3.116}$$

[17] This is now permitted since we have made $d - a$ very small.

The left-hand side is a function of E, the right-hand side is not. The left-hand side can assume values larger than one, the right-hand side cannot. Equation 3.116 can, therefore, be satisfied only for those values of E for which the left-hand side remains between -1 and $+1$.[18] In other words, in a crystal only certain energy bands are allowed for an electron. This result is borne out by a more complete theory and confirmed by experiments. Figure 3.9 shows a plot of

$$\phi(\alpha d) = \frac{Km}{\hbar\sqrt{2mE}} \sin\left(\frac{d}{\hbar}\sqrt{2mE}\right) + \cos\left(\frac{d}{\hbar}\sqrt{2mE}\right) \qquad (3.117)$$

as a function of αd, for various values of K.

The values of ϕ for which Eq. 3.116 can be satisfied are drawn heavily. They determine the so-called allowed energy bands of the crystal. Taking the allowed values of αd from Figure 3.9, we can calculate the allowed values of E from Eq. 3.100.

$$E = \frac{\hbar^2(\alpha d)^2}{2md^2} \qquad (3.118)$$

The result is shown in Figure 3.12. A look at Figures 3.11 and 3.12 shows what we would have expected. The larger K, that is the higher the potential separating the regions of zero potential, the narrower the energy bands. For $K \to \infty$ the crystal is reduced to a series of independent square wells, and the energy bands go over into the discrete eigenvalues that we had found for a square well in Chapter 3.1.

3.5   SUMMARY

Let us review our findings thus far. The transition from classical mechanics to quantum mechanics starts from the classical equation of motion of a system:

$$E = \frac{p^2}{2m} + V(r) \qquad (3.119)$$

The transition is accomplished in five steps.

(a) The total energy E is replaced by an operator

$$i\hbar \frac{\partial}{\partial t}$$

(b) The momentum p is replaced by an operator[19]

$$-i\hbar\nabla$$

---

[18] In a real crystal, N is very large, kd can vary almost continuously, and $\cos(kd)$ can assume any value between -1 and $+1$.

[19] This choice of sign is customary, since only the square of $i\hbar\nabla$ occurs in the Schrödinger equation, it does not make any difference *here*. We shall, however, come back to this point in Chapter 4.1.

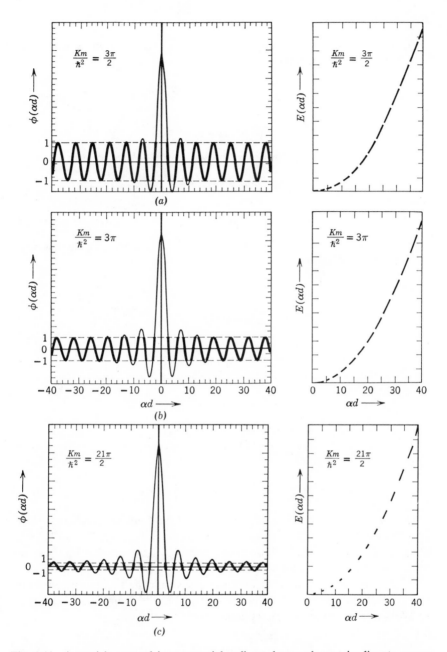

Fig. 3.11 A particle trapped in a potential well can have only certain discrete energy values. In a periodic potential these discrete energies are replaced by allowed energy bands.

Fig. 3.12 As the barriers between different crystal cells get bigger and bigger the energy bands shrink to the allowed energy values of the particle in an infinite square well.

(c) The resulting operator is applied to a wave function ψ and the resulting **time dependent Schrödinger equation**:

$$i\hbar \frac{\partial \psi}{\partial t} = -\frac{\hbar^2}{2m} \nabla^2 \psi + V(\mathbf{r})\psi \tag{3.120}$$

is solved for $\psi(\mathbf{r}, t)$.

(d)

$$\psi^*(\mathbf{r}, t)\psi(\mathbf{r}, t)\, d\tau$$

is the **probability** that the particle can be found in the volume element $d\tau$ at the time t.

If the potential does not contain an explicit time dependence, i.e., if

$$\frac{\partial V}{\partial t} = 0 \tag{3.121}$$

the Schrödinger equation (Eq. 3.120) is separable, and we obtain the **time independent Schrödinger equation**

$$\frac{-\hbar^2}{2m} \nabla^2 u(\mathbf{r}) + V(\mathbf{r})u(\mathbf{r}) = Eu(\mathbf{r}) \tag{3.122}$$

The eigenvalues of the **Hamiltonian** (operator)

$$H = -\frac{\hbar^2}{2m} \nabla^2 + V(\mathbf{r}) \tag{3.123}$$

are the possible **energy values** of the system. The square of the absolute value of the eigenfunction

$$u^*(\mathbf{r})u(\mathbf{r})$$

gives the **probability density** that the particle can be found at $\mathbf{r}$.

(e) The eigenfunctions $\psi(\mathbf{r}, t)$, or $u(\mathbf{r})$ for the time independent problem, must be **normalized** by forming

$$\int \psi^*(\mathbf{r}, t)\psi(\mathbf{r}, t)\, dt = 1 \quad \text{or} \quad \int u^*(\mathbf{r})u(\mathbf{r})\, d\tau = 1 \tag{3.124}$$

These integrations can only be performed if the **boundary conditions** of the problem are known. The boundary conditions together with Eq. 3.124 select the physically meaningful solutions of Eq. 3.120 or Eq. 3.122 from the many more mathematically possible ones.

It is obvious from the foregoing examples that the boundary conditions play a decisive role. The condition $u(\pm a) = 0$ in the example of the particle in a box eliminated all solutions that did not satisfy

$$E_n = \frac{n^2 \pi^2 \hbar^2}{8a^2 m}, \qquad n = 1, 2, 3, \ldots \tag{3.125}$$

Similarly the condition

$$\int u^*u\, dx = 1 \tag{3.126}$$

Fig. 3.13 In quantum mechanics, as well as in classical mechanics, the basic differential equations permit an infinite variety of solutions. Whether we hear music depends on the boundary conditions.

imposed on the solutions of the linear harmonic oscillator led to discrete energy eigenvalues. In the one-dimensional crystal the periodicity of the potential together with the condition

$$u(x + Nd) = u(x) \tag{3.127}$$

led to **allowed bands** of discrete energy eigenvalues. For finite values of N, Eq. 3.116 is, of course, satisfied only for a finite number of energy eigenvalues in each band.

We conclude[20] from our examples: Whenever a particle is confined to a certain region of space only certain discrete energy eigenvalues are allowed.

The great importance of the boundary conditions and the fact that confinement leads to discrete eigenvalues is by no means restricted to quantum mechanics. To illustrate this we mention two examples from classical physics.

Maxwells equations[21] encompass everything that is electromagnetic. Whether they describe the TV program on Channel 6 or the current generated in a hydroelectric power plant depends on the boundary conditions under which we solve them. The motion of a string can be described by a differential equation, and all the configurations the string can assume are compatible with this differential equation. If we impose boundary conditions, only certain configurations remain possible (Figure 3.13).

---

[20] A rigorous proof of this statement has not yet been found, although the widely varying conditions under which we have shown it to be valid make it plausible that the statement is universally true.

[21] J. R. Reitz and F. J. Milford, *Foundations of Electromagnetic Theory*, Addison-Wesley, Inc., Reading, Mass., 1960.

The Schrödinger equation is a linear differential equation or, to put it differently, the Hamiltonian operator is linear. According to the definition of linearity (see Appendix A1), this means that a superposition of solutions (or eigenfunctions) is also a solution. We have studied this for the case of the wave packet (see Chapter 2.2). We mention here in passing that a similar superposition is possible when different eigenfunctions belong to the same eigenvalue, a case to which we shall return later (Chapter 9).

We conclude this chapter with the introduction of some widely used expressions whose origins should by now be obvious to the reader:

A quantum mechanical system can usually exist in various states characterized by their wave functions. A state whose wave function is an eigenfunction of an operator is often called an **eigenstate** of that operator.

If the wave function of a state can be expressed by a linear combination of other wave functions we often refer to this as a **superposition of states.**

The labels (indices) of a set of discrete eigenvalues are often called the **quantum numbers** of the corresponding states.

Problems

3.1 A gas consisting of point particles of mass $m = 10^{-23} g$ is confined in a box whose volume is 1 cm³. According to the kinetic theory of gases the average velocity of the gas molecules is given by

$$\frac{mv^2}{2} = \tfrac{3}{2}kT$$

where $k = 1.38 \times 10^{-16}$ [erg/molecule °K] is Boltzmann's constant. Give a rough estimate of the temperature at which one can expect to see a deviation from the behavior predicted by the kinetic theory. Is this limitation of practical concern in the example given here?

3.2 (a) A proton or (b) an electron are trapped in a one-dimensional box of 10^{-8} cm length. What is the minimum energy these particles can have?

3.3 A particle is trapped in a one-dimensional box whose total length is L. Its energy is $E = \pi^2\hbar^2/2mL^2$. Where in the box is the particle most likely to be found?

3.4 An electron whose kinetic energy is 10 eV is trapped in a potential well whose walls are 10.01 V high. How thin must the walls of the well be in order for the electron to have an appreciable chance to escape?

3.5 A junction between two superconducting metals exhibits the a.c. Josephson effect. A voltage of 1.0 mV is applied across the junction. What is the frequency of the electromagnetic radiation emitted by the junction?

3.6 Determine the normalization factors B_0, B_1, and B_2 for the harmonic oscillator eigenfunctions of Table 3.1.

3.7 To find out why we have split off the factor $e^{-x^2/2}$ (Eq. 3.47), expand $u(x)$ itself into a series and substitute *this* series into Eq. 3.44). Compare the result with Eq. 3.56.

3.8 The energy eigenvalues of a molecule indicate that the molecule is a one-dimensional harmonic oscillator. In going from the first excited state to the ground state, the molecule emits a photon of $hv = 0.1$ eV. Assuming that the oscillating part of the molecule is a proton, calculate the probability that the proton is at a distance from the origin that would be forbidden to it by classical mechanics.

3.9 An electron is trapped in a one-dimensional harmonic oscillator potential. The frequency difference between the spectral lines emitted by the electron is $\Delta v = 10^{15}$ cycles/sec. The electron is in its ground state. What is the probability of finding the electron at a distance of $(+10^{-7} \pm 1 \times 10^{-9})$ cm from the origin? What is the probability of finding the electron at the center of the potential well in an interval of $\pm 10^{-9}$ cm?

3.10 Show that for infinitely large values of the constant K in Eq. 3.116 the allowed energy bands contract to points on the energy scale and that these points coincide with the energy eigenvalues of a particle in a one-dimensional box of total length d.

SOLUTIONS

3.1 A serious discrepancy with the kinetic theory of gases can certainly be expected if the average kinetic energy $mv^2/2 = \frac{3}{2}kT$ approaches the minimum energy of a particle in a box as given by Eq. 3.15:

$$E_1 = \frac{\pi^2 \hbar^2}{8a^2 m}$$

Since $a = \frac{1}{2}$ for a cubic box we let

$$\frac{\pi^2 \hbar^2}{2m} = \frac{3kT}{2} \quad \text{or} \quad T = \frac{\pi^2 \hbar^2}{3mk} = \frac{\pi^2 (6.626)^2 \cdot 10^{-54}}{4\pi^2 \cdot 3 \cdot 10^{-23} \cdot 1.38 \cdot 10^{-16}}$$

$$= 2.65 \cdot 10^{-15} \, {}^\circ \text{K}$$

This temperature is so low that quantum effects would be quite negligible even at the lowest temperatures that can be reached today ($\sim 10^{-3}$ °K). This does not mean that quantum effects cannot be observed in low temperature gases or liquids. In our example we had assumed point particles so that each particle has the full volume of the box available. Real gas molecules, of course, occupy a volume so that only a small intermolecular volume remains at very low temperatures. In that case we can easily observe quantum effects in liquid He at temperature of the order of 1°K.

3.8 If $E = hv = 0.1$ eV we have according to Table 3.1:

$$E_0 = \frac{\hbar\omega}{2} = 0.05$$

The force constant C is, according to Eq. 3.41:

$$C = m\omega^2$$

A classical harmonic oscillator with this force constant and a total energy E_0 has, according to Eq. 3.39, an amplitude of

$$A = \frac{2E_0}{c} = \frac{\hbar}{\sqrt{2mE_0}}$$

The ground state wave function is, according to Table 3.1,

$$u_0 = B_0 e^{-x^2/2} = \frac{1}{\sqrt{\sqrt{\pi}}} e^{-x^2/2}$$

using the normalization (Eq. 3.71). The probability of finding the proton outside the classically defined limit A is then

$$P(A) = 1 - \frac{1}{\sqrt{\pi}} \int_{-A}^{A} e^{-x^2}\, dx = 1 - \frac{2}{\sqrt{\pi}} \int_{0}^{A} e^{-x^2}\, dx$$

The reader may evaluate this integral using tables of the error integral

$$\frac{1}{\sqrt{2\pi}} \int_{0}^{t} e^{-x^2/2}\, dx$$

4

SOME THEOREMS AND DEFINITIONS

4.1 THE EXPECTATION VALUE, EHRENFEST'S THEOREM

The object of theoretical physics is to make quantitative statements about measurable parameters of a physical system. In classical physics there can be little doubt about the connection between a theoretical prediction and its experimental verification.

In the realm of quantum mechanics, the existence of the uncertainty principle forces us to reexamine this problem, raising the following questions.

(a) In what way can we extract a value for a dynamic variable from our theory?

(b) How is this value related to the outcome of an experiment designed to measure the variable?

Because of the uncertainty principle we cannot determine the parameters of a system with complete accuracy. The sad fact that experiments, for whatever reasons,[1] do not yield a unique result was well known to experimentalists before the advent of quantum mechanics, and they developed methods to deal with this regrettable situation. It is a common practice to repeat uncertain experiments several times and to determine the average of several measurements of, for example, a parameter x according to

$$\bar{x} = \frac{1}{n} \sum_{k=1}^{n} x_k \tag{4.1}$$

where x_k is the result of the kth measurement. If certain values x_k occur n_k times, we can write Eq. 4.1 as

$$\bar{x} = \frac{\sum\limits_{k=1}^{m} n_k x_k}{\sum\limits_{k=1}^{m} n_k} = \frac{1}{n} \sum_{k=1}^{m} n_k x_k \tag{4.2}$$

[1] We are thinking here of reasons as mundane as insufficient resolution, imperfect equipment, etc.

The probability that a measurement yields the value x_k is then

$$P_k = \frac{n_k}{\sum_{k=1}^{m} n_k} = \frac{n_k}{n} \tag{4.3}$$

Using this definition, we can write

$$\bar{x} = \sum_{k=1}^{m} P_k x_k \tag{4.4}$$

We shall follow the mathematical usage of calling the thus defined average, $\bar{x}$, the **expectation value** of x (often written as $\bar{x} = \langle x \rangle$). The definition (Eq. 4.4) can obviously be extended to a continuous distribution of probabilities, $P(x)$, by writing

$$\langle x \rangle = \int_{-\infty}^{\infty} P(x) x \, dx = \int_{-\infty}^{\infty} x \, dW \tag{4.5}$$

$dW = P(x) \, dx$ is the probability that x has a value between x and $x + dx$. Since the probability that x has any value at all must be one (a certainty), we find that $P(x)$ has to satisfy

$$\int_{-\infty}^{\infty} P(x) \, dx = 1 \tag{4.6}$$

$P(x)$ is usually called the **probability density** of x because it gives the probability per unit interval

$$P(x) = \frac{dW}{dx} \tag{4.7}$$

that x has a certain value. In forming the average of x according to Eq. 4.5, those values of x that are more probable contribute more heavily to the integral. These values are, so to speak, given more weight. For this reason, $\langle x \rangle$ as in Eq. 4.5 is sometimes called the **weighted average** of x, and $P(x)$ is also called the **weight function.** The procedure used to get the expectation value of a parameter x can be extended to functions of one or several parameters. We can define the expectation value of a function of, for example, three parameters $f(x, y, z)$ as

$$\langle f(x, y, z) \rangle = \iiint P(x, y, z) f(x, y, z) \, dx \, dy \, dz \tag{4.8}$$

In applying this to quantum mechanics, we remember that $u^*(\mathbf{r}) u(\mathbf{r}) \, d\tau$ is the probability that a particle can be found in the volume element $d\tau$. The expectation value $\langle \mathbf{r} \rangle$ of the coordinate vector of the particle is thus, according to Eq. 4.8,

$$\langle \mathbf{r} \rangle = \int u^*(\mathbf{r}) u(\mathbf{r}) \mathbf{r} \, d\tau \tag{4.9}$$

The expectation value, on the other hand, is defined as the average of many measurements. The connection between theory and experiment must, therefore, be given by

$$\langle \mathbf{r} \rangle = \int u^*(\mathbf{r})u(\mathbf{r})\mathbf{r} \, d\tau = \lim_{n \to \infty} \frac{1}{n} \sum_{k=1}^{n} \mathbf{r}_k \qquad (4.10)$$

where $\mathbf{r}_k$ is the result of the kth measurement of the particle coordinates. The only way to measure the coordinates of a particle is to perform a scattering experiment, i.e., to bounce some other particle off it. Obviously this will change the coordinates of the scattering particle and will, in general, make it unavailable for subsequent measurements. The way to get around this difficulty is to start with a large number of target particles in the same state and to bombard them with a stream of identical particles, thus averaging over a large number of measurements. Figure 4.1 shows a typical scattering experiment. The particle detector is moved around the target and measures the flux of scattered particles as a function of the angle θ. Symmetry considerations preclude a ϕ dependence as long as target particles and bombarding particles have their symmetry axes—if they have any—oriented at random. It might seem as if such a scattering experiment would yield nothing but $\langle \mathbf{r} \rangle$, the average location of *all* the target atoms, i.e., the center of mass of the target. This is not true. Among other things, we can obviously extract some information about the size of the scattering particles[2] from an analysis of the relative number of the scattered particles. This amounts to a measurement of $\langle |\Delta \mathbf{r}| \rangle$ and since $|\Delta \mathbf{r}|$ is the same for all the target particles it does not matter whether we measure

$$\langle |\Delta \mathbf{r}| \rangle = \frac{1}{n} \sum_{k=1}^{m} |\Delta \mathbf{r}_k| \qquad (4.11)$$

several times for the same particles or once for several identical particles.

Now we turn to the expectation value of the momentum. It is easily defined experimentally. We measure the momentum of a particle several times or we measure the momenta of several particles accelerated under identical conditions and take the average. But how can we define it theoretically? How do we determine the expectation value of an operator? Is it

$$\langle p_x \rangle = -\int i\hbar \frac{\partial}{\partial x} (\psi^* \psi) \, d\tau \qquad \text{or} \qquad \langle p_x \rangle = -\int \psi^* i\hbar \frac{\partial \psi}{\partial x} \, d\tau$$

or what? To find out, we follow a procedure given by Paul Ehrenfest.[3] Let

$$\langle x(t) \rangle = \int x \psi^* \psi \, d\tau \qquad (4.12)$$

---

[2] If the size of the bombarding particles is known.
[3] P. Ehrenfest, *Z. Physik* **45**, 455 (1927).

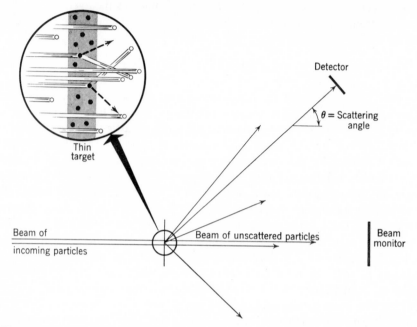

Fig. 4.1 In a typical scattering experiment a beam of identical incoming particles is scattered by a large number of identical target particles.

be the expectation value of the x coordinate of a particle moving along the x-axis.

$$v_x = \frac{d\langle x \rangle}{dt} \qquad (4.13)$$

must be the average velocity with which the wave packet (particle) moves. We conclude that[4]

$$m \frac{d\langle x \rangle}{dt} = \langle p_x \rangle \qquad (4.14)$$

should be the expectation value of the x-component of the particle momentum. We now proceed to express $\langle p_x \rangle$ in terms of the operator $i\hbar(\partial/\partial x)$.[5]

$$m \frac{d\langle x \rangle}{dt} = m \frac{d}{dt} \int x \psi^* \psi \, d\tau$$

$$= m \int x \left(\psi^* \frac{\partial \psi}{\partial t} + \psi \frac{\partial \psi^*}{\partial t} \right) d\tau \qquad (4.15)$$

[4] We restrict ourselves to $m(d\langle x \rangle/dt)$ instead of $m(d\langle \mathbf{r} \rangle/dt)$ purely for reasons of mathematical simplicity. The transition to three dimensions will be obvious for the final result.

[5] Note that x, y, and z are parameters of the integration, they are, therefore, independent of t, i.e., $dx/dt = dy/dt = dz/dt = 0$.

For $\partial\psi/\partial t$ and $\partial\psi*/\partial t$ we substitute expressions we obtain from the Schrödinger equation (Eq. 2.15) and its complex conjugate:

$$\frac{\partial\psi}{\partial t} = \frac{i\hbar}{2m}\nabla^2\psi - \frac{i}{\hbar}V(\mathbf{r})\psi \qquad (4.16)$$

$$\frac{\partial\psi*}{\partial t} = -\frac{i\hbar}{2m}\nabla^2\psi* + \frac{i}{\hbar}V(\mathbf{r})\psi* \qquad (4.17)$$

Substituting Eqs. 4.16 and 4.17 into Eq. 4.15, we obtain

$$\langle p_x \rangle = m\int \psi* x\,\frac{i\hbar}{2m}\nabla^2\psi\,d\tau - m\int \psi x\,\frac{i\hbar}{2m}\nabla^2\psi*\,d\tau \qquad (4.18)$$

Because of the hermiticity of the Laplace operator,[6] the second integral can be rewritten ·

$$\int x\psi\,\frac{i\hbar}{2m}\nabla^2\psi*\,d\tau = \int \psi*\,\frac{i\hbar}{2m}\nabla^2(x\psi)\,d\tau \qquad (4.19)$$

hence

$$\langle p_x \rangle = \frac{i\hbar}{2}\int \psi*[x\,\nabla^2\psi - \nabla^2(x\psi)]\,d\tau$$

$$= \frac{i\hbar}{2}\int \psi*(x\,\nabla^2\psi - 2\nabla x\,\nabla\psi - x\,\nabla^2\psi)\,d\tau$$

$$= -i\hbar\int \psi*\,\frac{\partial\psi}{\partial x}\,d\tau \qquad (4.20)$$

This answers our question. The expectation value of the momentum is obtained by sandwiching the momentum operator, $-i\hbar(\partial/\partial x)$, between the two eigenfunctions and integrating. Making the obvious extension to three dimensions, we get

$$\langle \mathbf{p} \rangle = \int \psi*(-i\hbar\,\nabla)\psi\,d\tau \qquad (4.21)$$

This equation is known as Ehrenfest's theorem. Now we can also explain the minus sign for which we had no reason earlier (see p. 56). The minus sign comes from an arbitrary choice we made in the phase of the wave function (Eq. 2.10). If, instead, we had used a wave function

$$\psi(x,t) = A\exp\left[-i\left(\frac{xp_x}{\hbar} - \frac{p_x^2 t}{2m\hbar}\right)\right] \qquad (4.22)$$

---

[6] See Appendix A.1.

it would have satisfied a Schrödinger equation

$$-\frac{\partial \psi}{\partial t} = \gamma \frac{\partial^2 \psi}{\partial x^2} \tag{4.23}$$

and this would have led to a plus sign in Ehrenfest's theorem. The choice of sign we have made here is universally accepted in the literature.

Equations 4.12 and 4.21 suggest a generalization of the averaging procedure they employ. Let $F(\mathbf{p}, \mathbf{r})$ be a function of the dynamic variables $\mathbf{p}$ and $\mathbf{r}$ describing some measurable property of a classical system. The expectation value of the same property in the corresponding quantum mechanical system can then be obtained in the following way: The function $F(\mathbf{p}, \mathbf{r})$ is replaced with the operator

$$F = F(-i\hbar\nabla, \mathbf{r}) \tag{4.24}$$

using the procedure given on p. 56. The expectation value $\langle F \rangle$ is found by letting

$$\langle F \rangle = \int \psi^* F \psi \, d\tau \tag{4.25}$$

where ψ is the wave function describing the system.

From all we have said thus far it should be clear that this statement is a *postulate* rather than a provable theorem. As such, it is in need of experimental verification which we provide by applying it to find the expectation value of the energy:

$$F(\mathbf{p}, \mathbf{r}) = \frac{p^2}{2m} + V(\mathbf{r}) \tag{4.26}$$

As we know, this translates into

$$F(-i\hbar\nabla, \mathbf{r}) = -\frac{\hbar^2}{2m}\nabla^2 + V(\mathbf{r}) = H \tag{4.27}$$

According to Eq. 2.22, we have

$$Hu = Eu \tag{4.28}$$

hence

$$\langle H \rangle = \int u^* Hu \, d\tau = E \int u^* u \, d\tau = E \tag{4.29}$$

This statement can be compared with experimental results, and we shall do this for the case of the hydrogen atom in a later chapter. This final postulate was introduced by Dirac and it completes the list of postulates on p. 57. The postulate introduced ad hoc at the end of Chapter 2.1 is now seen to be just a special case of Dirac's postulate.

4.2 DEGENERACY

Degeneracy is a phenomenon that we encounter in systems described by partial differential equations. We shall make ourselves familiar with it by contemplating a classical example. Figure 4.2 shows a circular membrane (drumhead), clamped around its periphery, in various states (modes) of vibration. The lines represent nodes, the circumference being a node in every

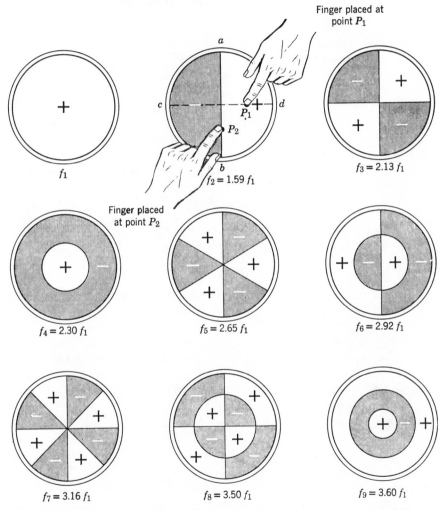

Fig. 4.2 The φ-dependent modes of vibration of a circular drumhead are twofold degenerate. The frequency in each of these modes is independent of the orientation of the nodes. This rotational symmetry can be lifted by an external perturbation.

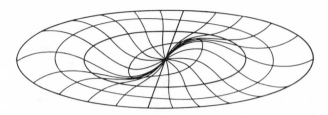

Fig. 4.3   A sketch of the drumhead vibrating in the mode f_6.

case. The plus and minus signs represent opposite displacements: at an instant when the plus areas are raised the minus areas are depressed. Figure 4.3 shows a perspective view of the drumhead vibrating in the mode f_6. The numbers underneath the individual figures give the frequency of the particular mode in multiples of the fundamental frequency f_1.

If we strike the membrane at the center we will obviously excite the first mode, the fourth mode, or one of the higher modes having complete rotational symmetry. We could also excite any combination of these modes. Let us assume that somehow we had managed to excite only the first mode. This mode is characterized completely by its eigenvalue, i.e., the frequency f_1 and the eigenfunction

$$A_1 = A_1(x, y, t) \tag{4.30}$$

that gives the deflection in the z-direction as a function of x, y, and t. From Figure 4.2a it is obvious that the x, y dependence of A is of the form

$$A_1(\sqrt{x^2 + y^2}, t) = A_1(r, t) \tag{4.31}$$

In other words, in a cylindrical coordinate system centered on the center of the membrane the eigenfunction does not depend on the azimuth φ. We turn now to the mode f_2. Here the eigenfunction depends on the angle φ. Obviously it will be possible to excite this mode in many different ways, depending on the azimuth φ at which we strike the initial blow. A blow at P_1, for example, would excite a vibration around the line a–b. A blow at P_2 would excite a vibration around the dotted line c–d. Because of the rotational symmetry of the membrane both vibrations must have the same frequency (eigenvalue). The two eigenfunctions, however, will be different. True, the second eigenfunction will result from the first one through a simple rotation of the coordinate system, *but* it cannot be written as a multiple of the first eigenfunction. An eigenvalue—here f_2—that has several eigenfunctions which are not linear combinations of each other,[7] is said to be degenerate.

[7] A linear combination of one function (as in our example) is, of course, simply a multiple of that function.

Inspection of Figure 4.2b makes it plausible that all possible vibrations of the mode f_2 can be described as a superposition of two vibrations: one around the axis a–b, the other around some other axis (for example, c–d[8]). This mode is, therefore, said to be twofold degenerate. In a more general case where we need at least n different eigenfunctions to describe all the possible modes of vibration we speak of n-fold degeneracy. Inspection of Figure 4.2 shows that the eigenfunctions of the vibrating membrane are, at most, twofold degenerate. Degeneracy is thus likely to occur in systems that have certain symmetries. Symmetry is, however, not the only cause of degeneracy. If the modes f_3 and f_7 (Figure 4.2), by some accident, had the same frequency (they do not), we would still call this a degeneracy even though the two eigenfunctions are not related by a simple symmetry operation.

Degeneracy of an eigenvalue is much more than a mathematical curiosity. Only a few quantum mechanical problems can be solved directly. Usually we have to solve a simpler problem and then apply perturbation theory to take into account all the interactions present. In the example of Figure 4.2, it may not matter whether the membrane vibrates around a–b or c–d since the frequency is the same in either case. However, if the symmetry is disturbed by a perturbation (for example, a finger placed at the point P_1), the two modes will have different frequencies and the degeneracy is lifted. Thus the concept of degeneracy will become important later when we discuss perturbation theory. At the moment we mention only that the fine structure and hyperfine structure of spectral lines result from degeneracies that are lifted by small perturbations.

PROBLEMS

4.1 Determine the expectation value of the kinetic energy and the potential energy of the linear harmonic oscillator in the states with $n = 0$, $n = 1$, and $n = 2$.

4.2 Determine the expectation value of x and p_x of the linear harmonic oscillator for the states with $n = 0$, $n = 1$, $n = 2$. Interpret your result.

4.3 The wave function of a particle is

$$u(x) = A \exp\left[i\left(\alpha x - \frac{\alpha^2 \hbar t}{2m}\right)\right]$$

What is the expectation value of its momentum?

[8] A detailed theoretical analysis bears this out. The eigenfunctions of the membrane are of the form $\sin (n\varphi)I_n$ and $\cos (n\varphi)I_n$ where the I_n are Bessel functions. Obviously any eigenfunction $\sin (n\varphi + \alpha)I_n$ can be written as a linear combination:

$$\sin (n\varphi + \alpha) = [a \sin (n\varphi) + b \cos (n\varphi)]I_n$$

4.4 Derive an expression for the restoring force in a linear harmonic oscillator. (*Hint.* Since the force constantly changes direction and the system is symmetrical about the origin, the expectation value of the force, i.e., its average, is of course equal to zero. A useful estimate can nevertheless be obtained by considering the average of the absolute value of the distance x from the origin.)

4.5 Derive an expression for the expectation value of the z-component of the angular momentum. (*Hint.* Classically the angular momentum is given by $\mathbf{L} = \mathbf{r} \times \mathbf{p}$.)

Solution of 4.3. The expectation value of the momentum is

$$\langle p \rangle = \int_{-\infty}^{\infty} u^* \left(-i\hbar \frac{\partial u}{\partial x} \right) dx$$

Now

$$-i\hbar \frac{\partial u}{\partial x} = -Ai\hbar \cdot i\alpha \cdot \exp\left[i\left(\alpha x - \frac{\alpha^2 \hbar t}{2m} \right) \right] = \alpha \hbar u$$

hence

$$\langle p \rangle = \alpha \hbar \int_{-\infty}^{\infty} u^* u \, dx = \alpha \hbar$$

5

THE HYDROGEN ATOM

5.1 THE HAMILTONIAN OF THE HYDROGEN ATOM IN THE CENTER OF MASS SYSTEM

A hydrogen atom consists of a proton and an electron, held together by the electrostatic attraction the two hold for each other. Classical physics would draw the following picture of such a system (Figure 5.1). The total energy of the proton and electron in this coordinate system is given by

$$E = \frac{p_e^2}{2m_e} + \frac{p_p^2}{2m_p} - \frac{e^2}{|\mathbf{r}_e - \mathbf{r}_p|} \tag{5.1}$$

total energy = kinetic energy + kinetic energy + potential energy of
          of the electron   of the proton   proton and electron due
                                        to their mutual coulomb
                                        attraction[1]

where

$\mathbf{p}_e$ = momentum of the electron, $\mathbf{p}_p$ = momentum m_e = electron mass
                                  of the proton

m_p = proton mass $|\mathbf{r}_e - \mathbf{r}_p|$ = distance between electron and proton.

---

[1] The validity of Coulomb's law has been established experimentally with extreme accuracy. *Plimpton* and *Lawton* have shown, in 1936, that the exponent of *r* in Coulomb's law is $2 \pm 2 \times 10^{-9}$. Their measurement was done, however, using macroscopic laboratory equipment. In using Coulomb's law in Eqs. 5.1 and 5.2, we make the bold assumption that it will still be valid in the realm of the submicroscopic where other well-known laws (those of classical mechanics) fail. The agreement between the quantum mechanical theory of the hydrogen atom and the experimental results, which we are about to establish, will prove at the same time that Coulomb's law is valid down to distances of the order of 10^{-8} cm.

It might be mentioned at this point that the validity of Coulomb's law for protons has been established down to $\sim 10^{-13}$ cm. Electrons, to the best of our knowledge, behave like true point particles, obeying Coulomb's law as far as we can measure.

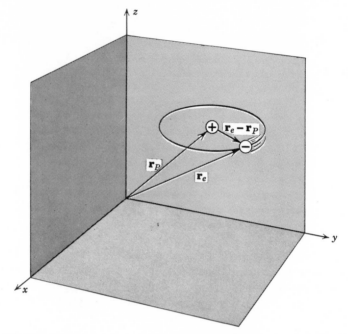

Fig. 5.1 The classical picture of a hydrogen atom as described in an arbitrary cartesian coordinate system.

Using the substitutions of p. 19, we obtain the following Hamiltonian:

$$H = -\frac{\hbar^2}{2m_e}\nabla_e^{\;2} - \frac{\hbar^2}{2m_p}\nabla_p^{\;2} - \frac{e^2}{|\mathbf{r}_e - \mathbf{r}_p|} \tag{5.2}$$

The operator $\nabla_e^{\;2}$ operates only on the electron coordinates; the operator $\nabla_p^{\;2}$ operates only on the proton coordinates.

This Hamiltonian describes the system completely, but it tells us more than we care to know. It describes not only the behavior of the electron and proton due to the coulomb interaction but also any motion the atom as a whole might make. In classical physics we usually distinguish between these two parts of the motion by transforming to a-center-of-mass coordinate system. In this way we obtain two equations of motion, one describing the motion of the center of mass in some "laboratory coordinate system," the other describing the motion of the parts of the system relative to their center of mass. We use the same technique here and introduce the coordinate vector of the center of mass:

$$\mathbf{R} = \begin{pmatrix} X \\ Y \\ Z \end{pmatrix} = \frac{m_e\mathbf{r}_e + m_p\mathbf{r}_p}{m_e + m_p} \tag{5.3}$$

Since the coulomb force between the proton and the electron depends only on their distance from each other, we need only to introduce one other vector to describe the system completely

$$\mathbf{r} = \begin{pmatrix} x \\ y \\ z \end{pmatrix} = \mathbf{r}_e - \mathbf{r}_p \tag{5.4}$$

To carry out the transformation of Eq. 5.2 to the center-of-mass coordinate system we form

$$\frac{\partial \psi}{\partial x_e} = \frac{\partial \psi}{\partial X}\frac{\partial X}{\partial x_e} + \frac{\partial \psi}{\partial x}\frac{\partial x}{\partial x_e} \tag{5.5}$$

Using Eqs. 5.3 and 5.4 this can be written as

$$\frac{\partial \psi}{\partial x_e} = \frac{m_e}{m_e + m_p}\frac{\partial \psi}{\partial X} + \frac{\partial \psi}{\partial x}, \tag{5.6}$$

Similarly

$$\frac{\partial \psi}{\partial x_p} = \frac{m_p}{m_e + m_p}\frac{\partial \psi}{\partial X} - \frac{\partial \psi}{\partial x}. \tag{5.7}$$

Analogous expressions are obtained for

$$\frac{\partial \psi}{\partial y_e}, \frac{\partial \psi}{\partial z_e}, \frac{\partial \psi}{\partial y_p}, \quad \text{and} \quad \frac{\partial \psi}{\partial z_p}$$

Introducing the **reduced mass**

$$m = \frac{m_e m_p}{m_e + m_p} \tag{5.8}$$

we obtain

$$\nabla_e = \frac{m}{m_p}\nabla_R + \nabla \tag{5.9}$$

and

$$\nabla_p = \frac{m}{m_e}\nabla_R - \nabla \tag{5.10}$$

where ∇_R operates only on the coordinates of the center of mass and ∇ only on the coordinates x, y, z. The Hamiltonian becomes, therefore,

$$H = \frac{-\hbar^2}{2m_e}\left(\frac{m}{m_p}\nabla_R + \nabla\right)^2 - \frac{\hbar^2}{2m_p}\left(\frac{m}{m_e}\nabla_R - \nabla\right)^2 - \frac{e^2}{r} \tag{5.11}$$

Keeping in mind that Eq. 5.11 is an operator equation, we carry out the squares and obtain

$$H = \frac{-\hbar^2}{2m_e}\left[\frac{m^2}{m_p^2}\nabla^2_R + \frac{m}{m_p}(\nabla_R\nabla + \nabla\nabla_R) + \nabla^2\right]$$
$$- \frac{\hbar^2}{2m_p}\left[\frac{m^2}{m_e^2}\nabla^2_R - \frac{m}{m_e}(\nabla_R\nabla + \nabla\nabla_R) + \nabla^2\right] - \frac{e^2}{r} \tag{5.12}$$

or

$$H = \frac{-\hbar^2}{2(m_e + m_p)}\nabla^2_R - \frac{\hbar^2}{2m}\nabla^2 - \frac{e^2}{r} \tag{5.13}$$

This Hamiltonian is the sum of a part $-\hbar^2/[2(m_e + m_p)]\nabla^2_R$ that depends only on R and a part $-[(\hbar^2/2m)\nabla^2 + e^2/r]$ that depends only on r. It is easy to show that if this is the case, we can separate the Schrödinger equation by substituting

$$\psi(\mathbf{r}, \mathbf{R}) = \phi(\mathbf{R})u(r) \tag{5.14}$$

The two parts of the eigenfunction, $\phi(\mathbf{R})$ and $u(r)$, are connected only through the common separation constant. In this case, the separation constant contains the kinetic energy of the center-of-mass of the atom. Unless the atom is placed in a box, this kinetic energy can assume any value and, as a result, the eigenfunctions $\phi(\mathbf{R})$ and $u(r)$ are independent of each other.[2] Physically this means, of course, that the internal state of the atom is independent of the motion of its center of mass.

To carry out the separation, we substitute Eq. 5.14 into Eq. 5.13, obtaining

$$\frac{-\hbar^2}{2(m_e + m_p)}u\nabla^2_R\phi - \frac{\hbar^2}{2m}\phi\nabla^2 u - \frac{e^2}{r}u\phi = E_t u\phi \tag{5.15}$$

We divide by $u\phi$

$$\frac{-\hbar^2}{2(m_e + m_p)\phi}\nabla^2_R\phi - \frac{\hbar^2}{2mu}\nabla^2 u - \frac{e^2}{r} = E_t \tag{5.16}$$

We group together the r and the R dependent terms and equate them to the same constant:

$$\frac{-\hbar^2}{2(m_e + m_p)\phi}\nabla^2_R\phi = E_c = \frac{\hbar^2}{2mu}\nabla^2 u + \frac{e^2}{r} + E_t \tag{5.17}$$

or

$$\frac{-\hbar^2}{2(m_e + m_p)}\nabla^2_R\phi = E_c\phi \tag{5.18}$$

[2] Later in this chapter we shall encounter situations where the separation constant, itself, is quantized. In this case the two factors of the eigenfunction will only be conditionally independent.

and, using

$$E = E_t - E_c: \tag{5.19}$$

$$\frac{-\hbar^2}{2m} \nabla^2 u - \frac{e^2 u}{r} = Eu \tag{5.20}$$

The solution of Eq. 5.18 is a plane wave, obviously describing whatever motion the center of mass, i.e., the entire atom makes. This part is of no interest to us. Equation 5.20 is the Schrödinger equation of a particle moving in a fixed potential (see Eq. 2.28) except that the electron mass has been replaced with the reduced mass. For this reason the separation in the center-of-mass system is sometimes referred to as the reduction of a two-body problem to a single-body problem.

5.2 SEPARATION OF THE SCHRÖDINGER EQUATION IN SPHERICAL POLAR COORDINATES

Since the potential of the hydrogen atom has spherical symmetry it is appropriate to transform to spherical polar coordinates before we attempt a solution. The coordinate system we shall use is shown in Figure 5.2. The Laplace operator

$$\nabla^2 = \frac{\partial^2}{\partial x^2} + \frac{\partial^2}{\partial y^2} + \frac{\partial^2}{\partial z^2}$$

becomes in these coordinates[3]

$$\nabla^2 = \frac{1}{r^2} \frac{\partial}{\partial r} \left(r^2 \frac{\partial}{\partial r} \right) + \frac{1}{r^2 \sin \vartheta} \frac{\partial}{\partial \vartheta} \left(\sin \vartheta \frac{\partial}{\partial \vartheta} \right) + \frac{1}{r^2 \sin^2 \vartheta} \frac{\partial^2}{\partial \varphi^2} \tag{5.21}$$

Using Eq. 5.21 in Eq. 5.20 yields

$$\frac{1}{r^2} \frac{\partial}{\partial r} \left(r^2 \frac{\partial u}{\partial r} \right) + \frac{1}{r^2 \sin \vartheta} \frac{\partial}{\partial \vartheta} \left(\sin \vartheta \frac{\partial u}{\partial \vartheta} \right)$$

$$+ \frac{1}{r^2 \sin^2 \vartheta} \frac{\partial^2 u}{\partial \varphi^2} + \frac{2m}{\hbar^2} \left(\frac{e^2}{r} + E \right) u = 0 \tag{5.22}$$

We try to separate this, substituting[4]

$$u(\mathbf{r}) = \chi(r) Y(\vartheta, \varphi) \tag{5.23}$$

$$Y \frac{\partial}{\partial r} \left(r^2 \frac{\partial \chi}{\partial r} \right) + \frac{\chi}{\sin \vartheta} \frac{\partial}{\partial \vartheta} \left(\sin \vartheta \frac{\partial Y}{\partial \vartheta} \right) + \frac{\chi}{\sin^2 \vartheta} \frac{\partial^2 Y}{\partial \varphi^2}$$

$$+ \frac{2mr^2}{\hbar^2} \left(\frac{e^2}{r} + E \right) \chi Y = 0 \tag{5.24}$$

---

[3] See for example, W. Kaplan *Advanced Calculus*, Addison-Wesley Publishing Co., Inc., Reading, Mass. 1953.
[4] We shall discuss the physical significance of this separation at the end of Chapter 5.9.

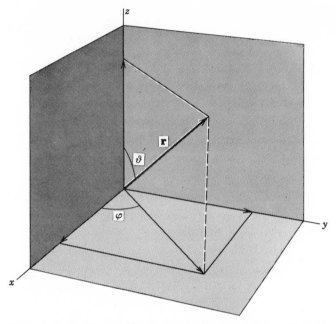

Fig. 5.2 The spherical polar coordinates used in the description of the hydrogen atom.

If we divide by χY, this separates into an r-dependent equation and a ϑ- and φ-dependent equation:

$$\frac{d}{dr}\left(r^2 \frac{d\chi}{dr}\right) + \frac{2mr^2}{\hbar^2}\left(\frac{e^2}{r} + E\right)\chi = A\chi \tag{5.25}$$

and

$$\frac{1}{\sin\vartheta}\frac{\partial}{\partial\vartheta}\left(\sin\vartheta \frac{\partial Y}{\partial\vartheta}\right) + \frac{1}{\sin^2\vartheta}\frac{\partial^2}{\partial\varphi^2}Y = -AY \tag{5.26}$$

where A is the separation constant. Now we note that Eq. 5.26 is independent of the total energy E and the potential energy V except for the connection with Eq. 5.25 by way of the separation constant A. This means that Eq. 5.26 is valid for any central potential, $V = V(r)$, and any value of the total energy E. This does not imply that the solutions of Eq. 5.26 are independent of V or E since the separation constant A is common to Eq. 5.26 and the E and V dependent Eq. 5.25. Physically meaningful solutions of Eq. 5.26 must, of course, have values of A that are compatible with solutions of Eq. 5.25 and vice versa. We postpone the solution of Eq. 5.25 and focus our attention on Eq. 5.26.

5.3   THE SEPARATION OF THE ANGULAR WAVE EQUATION

Since separation of the variables has served us well thus far, we try it again by substituting

$$Y(\vartheta, \varphi) = P(\vartheta)\phi(\varphi) \tag{5.27}$$

into Eq. 5.26. The result is

$$\frac{\sin \vartheta}{P} \frac{\partial}{\partial \vartheta} \left(\sin \vartheta \frac{\partial P}{\partial \vartheta} \right) + A \sin^2 \vartheta = -\frac{1}{\phi} \frac{\partial^2 \phi}{\partial \varphi^2} \tag{5.28}$$

If we name the separation constant m^2, as everybody does, this separates into

$$\sin \vartheta \frac{d}{d\vartheta} \left(\sin \vartheta \frac{dP}{d\vartheta} \right) + AP \sin^2 \vartheta = m^2 P \tag{5.29}$$

and

$$-m^2 \phi = \frac{d^2 \phi}{d\varphi^2} \tag{5.30}$$

This constant has nothing to do with the reduced mass m introduced earlier. Equation 5.30 can be integrated immediately to yield

$$\phi = \phi_0 e^{\pm im\varphi} \tag{5.31}$$

Since the wave function should not change if we change φ by 2π (thus coming back to the same point) we have to require that

$$\phi(\varphi) \equiv \phi(\varphi + 2\pi) \equiv \phi_0 e^{\pm im\varphi} \cdot e^{\pm im 2\pi} = \phi_0 e^{\pm im\varphi} \tag{5.32}$$

This will be the case if, and only if, m is an integer. We have just found the first quantum number of the hydrogen atom:

We call it the **magnetic quantum number** for reasons that will become apparent later. It should be noted here that the physical meaning of m is not evident from anything we have said thus far. To solve Eq. 5.29, we make the substitutions:

$$\xi = \cos \vartheta; \qquad \frac{d}{d\vartheta} = -\sin \vartheta \frac{d}{d\xi} \tag{5.33}$$

Hence

$$-\sin^2 \vartheta \frac{d}{d\xi} \left(-\sin^2 \vartheta \frac{dP}{d\xi} \right) + AP \sin^2 \vartheta = m^2 P \tag{5.34}$$

or

$$(1 - \xi^2) \frac{d}{d\xi} \left((1 - \xi^2) \frac{dP}{d\xi} \right) + A(1 - \xi^2)P = m^2 P \tag{5.35}$$

If we change the name of the constant in Eq. 5.35 to

$$A = l(l + 1) \tag{5.36}$$

Equation 5.35 is recognized as the well-known **associated Legendre equation**

$$(1 - \xi^2)\frac{d}{d\xi}\left((1 - \xi^2)\frac{dP}{d\xi}\right) + l(l + 1)(1 - \xi^2)P = m^2 P \tag{5.37}$$

Its solutions are known as the associated Legendre functions. The reader who is familiar enough with spherical harmonics to have recognized the Schrödinger equation (Eq. 5.26) as the differential equation of the spherical harmonics may at once advance to p. 83 where the solutions of Eq. 5.26 are listed and discussed. The reader who lacks mathematical proficiency is well advised to bear with us through the following purely mathematical discussion. It will lead to the derivation of the spherical harmonics: a class of functions that plays a decisive role not only in quantum mechanics but also in the theory of electricity and magnetism.

5.4 LEGENDRE POLYNOMIALS

A look at Eqs. 5.30 and 5.35 shows us that one particular value of m simplifies both equations: $m = 0$. We, therefore, try to solve Eq. 5.35 for this special case.

$$\frac{d}{d\xi}\left((1 - \xi^2)\frac{dP}{d\xi}\right) + AP = 0 \tag{5.38}$$

We substitute[5]

$$(a) \quad P = \sum_0 a_k \xi^k \quad \text{and} \quad (b) \quad \frac{dP}{d\xi} = \sum_1 k a_k \xi^{k-1} \tag{5.39}$$

into Eq. 5.38[6]

$$\sum_0 (k + 1)(k + 2)a_{k+2}\xi^k - \sum_0 k(k + 1)a_k \xi^k + A\sum_0 a_k \xi^k \equiv 0 \tag{5.40}$$

Equation 5.40 is an identity if, and only if, all the coefficients of equal powers of ξ vanish separately, i.e., if

$$a_{k+2} = a_k \frac{k(k + 1) - A}{(k + 1)(k + 2)} \tag{5.41}$$

Now

$$\lim_{k \to \infty} \frac{a_{k+2}}{a_k} = 1 \tag{5.42}$$

---

[5] For reasons soon to become apparent we are vague about the upper limit of these sums.
[6] The indices under a sum can, of course, be renamed freely, i.e.,

$$\sum_1 k(k + 1)a_{k+1}\xi^{k-1} \equiv \sum_0 (k + 1)(k + 2)a_{k+2}\xi^k$$

This means that Eq. 5.39a is[7] divergent for $\xi = 1$ (remember $\xi = \cos \vartheta$). At this point we have to invoke the boundary conditions of the hydrogen problem to select from the many mathematically possible solutions of Eq. 5.38 those that are physically meaningful. Since the attractive force between the electron and the proton abates gradually as r increases, there is no well-defined boundary. Therefore we have to settle for the condition

$$\int u^*u \, d\tau = 1 \tag{5.43}$$

which states merely that there is one electron and that it is somewhere. Since we had shown that the Schrödinger equation is separable, we could write

$$u = \chi(r)P(\vartheta)\phi(\varphi) \tag{5.44}$$

The three functions χ, P, and ϕ are related only through the common separation constants m^2 and $A = l(l + 1)$. We cannot, therefore, rely on one to make up for infinities in the others and must insist that

$$\int \chi^*\chi r^2 \, dr, \qquad \int P^*P \sin \vartheta \, d\vartheta \qquad \text{and} \qquad \int \phi^*\phi \, d\varphi$$

all remain finite. Hence, unless

$$\int P^*(\vartheta)P(\vartheta) \sin \vartheta \, d\vartheta \tag{5.45}$$

and, therefore, $P(\vartheta)$ remains finite[8], Eq. 5.43 cannot be satisfied. This means in view of Eq. 5.42 that Eq. 5.39a has to break off after a finite number of terms and that for some $k = l$

$$k(k + 1) = l(l + 1) = A \qquad l = 0, 1, 2, 3, \ldots \tag{5.46}$$

We have thus found another quantum number for the hydrogen atom

Again, we emphasize that nothing said thus far makes the physical meaning of l obvious. The polynomials

$$P_l = \sum_0^l a_k \xi^k \tag{5.47}$$

---

[7] Or rather could be. The criterion (Eq. 5.42) gives what is known as the *ambiguous case*, allowing either convergence or divergence.

[8] The fact that the integral (Eq. 5.45) is finite does not necessarily imply that $P(\vartheta)$, itself, is finite. However, for the kind of smoothly varying functions we are dealing with in physical applications, one condition always implies the other.

are known as **Legendre polynomials.** Since Eq. 5.38 determines P only up to a constant factor, and since the P_l—being polynomials—have no singularities for $-1 \leqslant \xi \leqslant 1$, they can be normalized. It is customary to normalize the Legendre polynomials by requiring

$$P_l(1) = 1 \qquad (5.48)$$

This together with Eq. 5.38 is sufficient to determine the coefficients of the Legendre polynomials unambiguously.

5.5 THE ASSOCIATED LEGENDRE FUNCTIONS

We return to the associated Legendre equation (Eq. 5.37) and make the substitution:

$$P = (\xi^2 - 1)^{m/2} w(\xi) \qquad (5.49)$$

This leads to

$$\frac{dP}{d\xi} = m\xi(\xi^2 - 1)^{(m/2)-1} w(\xi) + (\xi^2 - 1)^{m/2} \frac{dw}{d\xi} \qquad (5.50)$$

and

$$\frac{d^2P}{d\xi^2} = m(m - 2)\xi^2(\xi^2 - 1)^{(m/2)-2} w + m(\xi^2 - 1)^{(m/2)-1} w$$

$$+ m(\xi^2 - 1)^{(m/2)-1} \frac{dw}{d\xi} + m(\xi^2 - 1)^{(m/2)-1} \frac{dw}{d\xi} + (\xi^2 - 1)^{(m/2)} \frac{d^2w}{d\xi^2}$$

$$= (\xi^2 - 1)^{m/2} \Big\{ w[m(m - 2)\xi^2(\xi^2 - 1)^{-2} + m(\xi^2 - 1)^{-1}]$$

$$+ \frac{dw}{d\xi} 2m\xi(\xi^2 - 1)^{-1} + \frac{d^2w}{d\xi^2} \Big\} \qquad (5.51)$$

Substituting Eqs. 5.49, 5.50, and 5.51 into Eq. 5.38 yields—after collecting terms of equal order:

$$(\xi^2 - 1) \frac{d^2w}{d\xi^2} + (m + 1)2\xi \frac{dw}{d\xi} - [l(l + 1) - m(m + 1)]w = 0 \quad (5.52)$$

We return to the Legendre equation (Eq. 5.38).

$$\frac{d}{d\xi} \Big[(1 - \xi^2) \frac{dP_l}{d\xi} \Big] + l(l + 1)P_l = 0 \qquad (5.53)$$

Remembering Leibnitz's rule

$$(uv)^{(m)} = u^{(m)}v + \cdots + \frac{m(m - 1)}{2!} u^{(2)}v^{(m-2)} + mu^{(1)}v^{(m-1)} + uv^{(m)} \quad (5.54)$$

we differentiate Eq. 5.38 m-times, letting

$$(1 - \xi^2) = u \qquad \text{and} \qquad \frac{dP}{d\xi} = v \tag{5.55}$$

$$-m(m + 1)\frac{d^m P_l}{d\xi^m} - (m + 1)2\xi\frac{d^{m+1}}{d\xi^{m+1}}P_l$$

$$+ (1 - \xi^2)\frac{d^{m+2}}{d\xi^{m+2}}P_l + l(l + 1)\frac{d^m}{d\xi^m}P_l = 0 \tag{5.56}$$

Collecting terms of equal order yields

$$(\xi^2 - 1)\frac{d^2}{d\xi^2}\left(\frac{d^m P_l}{d\xi^m}\right) + 2(m + 1)\frac{d}{d\xi}\left(\frac{d^m P_l}{d\xi^m}\right)$$

$$- [l(l + 1) - m(m + 1)]\frac{d^m P_l}{d\xi^m} = 0 \tag{5.57}$$

We compare Eq. 5.57 with Eq. 5.52 and find them to be identical if we let

$$w = \frac{d^m P_l}{d\xi^m} \tag{5.58}$$

We sum up. The P_l are solutions of the Legendre equation (Eq. 5.38). $d^m P_l/d\xi^m$ satisfies Eq. 5.57, derived from Eq. 5.38 through m-fold differentiation. Equation 5.57 is none other than Eq. 5.52 which was obtained by substituting

$$(\xi^2 - 1)^{m/2}\frac{d^m P_l}{d\xi^m} \tag{5.59}$$

into the associated Legendre equation (Eq. 5.37). The solutions of the associated Legendre equation (Eq. 5.37) must, therefore, be

$$P_l{}^m = (\xi^2 - 1)^{m/2}\frac{d^m P_l}{d\xi^m} \tag{5.60}$$

These functions are called the **associated Legendre functions.** Since the P_l are polynomials of the order l, we cannot differentiate them more than l times. We must, therefore, insist that

$$m \leqslant l \tag{5.61}$$

We shall see later that this inequality has a very evident physical meaning. In view of Eqs. 5.27, 5.31, and 5.60, we can, therefore, state: The angular part of the Schrödinger equation for any central potential[9] is solved by the **spherical harmonics:**

$$Y_{lm} = N_{lm}e^{\pm im\varphi}P_l{}^m(\vartheta) \tag{5.62}$$

[9] In other words, for any potential that depends only on the magnitude of **r**.

where the N_{lm} are normalization constants. The name spherical harmonics suggests that the Y_{lm} are some kind of three-dimensional equivalent of the trigonometric functions. This is, indeed, true. The Y_{lm} are not only periodic in φ, and ϑ, but they also satisfy

$$\int Y^*_{lm} Y_{kn} \sin \vartheta \, d\vartheta = 0 \tag{5.63}$$

for $l \neq k$ or $m \neq n$ in obvious analogy to Eqs. A.2.7 and A.2.8. A set of functions that satisfies conditions like Eq. 5.63 or Eqs. A.2.7 and A.2.8 is said to be **orthogonal.** If the functions can also be normalized, as the Y_{lm} and the trigonometric functions obviously can, they are said to be **orthonormal.** We shall forgo the proof of the orthogonality of the spherical harmonics and return to it later (Chapter 6.1) in connection with a more general discussion of orthonormal sets.

5.6 THE SOLUTION OF THE RADIAL PART OF THE SCHRÖDINGER EQUATION

We return to Eq. 5.25, substituting for A the value we had found while solving the Legendre equation.

$$\frac{d}{dr} \left(r^2 \frac{d\chi}{dr} \right) + \frac{2mr^2}{\hbar^2} \left(\frac{e^2}{r} + E \right) \chi = l(l+1)\chi \tag{5.64}$$

Just as in the case of the harmonic oscillator (Chapter 3.3), substitution of a power series into Eq. 5.25 would lead to a recurrence relation connecting three coefficients with each other. Again, we have to resort to some mathematical slight of hand to avoid this. We substitute

$$\eta(r) = r\chi(r) \tag{5.65}$$

into Eq. 5.25. Since

$$\frac{d\chi}{dr} = \frac{1}{r} \frac{d\eta}{dr} - \frac{\chi}{r} = \frac{1}{r} \frac{d\eta}{dr} - \frac{\eta}{r^2} \tag{5.66}$$

this substitution yields the following differential equation for $\eta(r)$:

$$\frac{d^2\eta}{dr^2} + \frac{2m}{\hbar^2} \left(\frac{e^2}{r} + E \right) \eta = l(l+1) \frac{\eta}{r^2} \tag{5.67}$$

This would still lead to a useless recurrence relation, and we make one more substitution:

$$\eta(r) = \zeta(r) e^{-\alpha r} \tag{5.68}$$

This yields

$$\frac{d^2\zeta}{\partial r^2} - 2\alpha \frac{d\zeta}{dr} + \left[\frac{2me^2}{\hbar^2 r} + \frac{2mE}{\hbar^2} + \alpha^2 - \frac{l(l+1)}{r^2} \right] \zeta = 0 \tag{5.69}$$

The value of α in Eq. 5.68 was left open, and nothing can stop us from letting

$$\alpha^2 = -\frac{2mE}{\hbar^2} \tag{5.70}$$

Hence

$$\frac{d^2\zeta}{dr^2} - 2\sqrt{\frac{-2mE}{\hbar^2}}\frac{d\zeta}{dr} + \left(\frac{2me^2}{\hbar^2 r} - \frac{l(l+1)}{r^2}\right)\zeta = 0 \tag{5.71}$$

A final substitution

$$\zeta = \sum_0 c_k r^k \tag{5.72}$$

yields the recurrence relation:

$$c_{k+1} = \frac{2\left(k\sqrt{\dfrac{-2mE}{\hbar^2}} - \dfrac{me^2}{\hbar^2}\right)}{k(k+1) - l(l+1)} c_k \tag{5.73}$$

At this point we have to invoke the boundary conditions. If the hydrogen atom exists, the electron must be bound, and we have to insist that

$$\int \chi^*(r)\chi(r)r^2\, dr$$

remains finite.[10] Hence

$$\int \chi^*\chi r^2\, dr = \int \eta^*\eta\, dr = \int \zeta^2 e^{-2\alpha r}\, dr \leqslant M \tag{5.74}$$

where M is some finite number. This means that for large r

$$\zeta = \sum c_k r^k \tag{5.75}$$

has to converge faster than $e^{\alpha r}$ in order for Eq. 5.74 to remain finite. Now, the ratio of two coefficients c_k becomes for large k

$$\lim_{k\to\infty}\frac{c_{k+1}}{c_k} = \frac{2\alpha}{k} \tag{5.76}$$

On the other hand, if we expand $e^{\alpha r}$, the ratio of two consecutive coefficients tends towards α/k for large values of k. Hence Eq. 5.74 can be satisfied only if the series Eq. 5.72 breaks off after a finite number of terms. This is possible only if in Eq. 5.73 for some integer $k = n$:

$$\frac{-n^2 2mE}{\hbar^2} = \frac{m^2 e^4}{\hbar^4} \tag{5.77}$$

or

$$E = -\frac{me^4}{2n^2\hbar^2} \tag{5.78}$$

[10] See the pertinent remarks in Chapter 5.4.

This quantum number is the last of the three quantum numbers hidden behind the Schrödinger equation of the hydrogen atom. It is called the **principal quantum number** or the quantum number of the total energy.

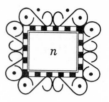

We conclude as follows.

(a) The total energy of a hydrogen atom is negative. This is not startling. In writing Eq. 5.1, we had fixed the energy scale—as is customary for systems with finite binding energy—so that the total energy was zero for a proton and an electron resting at a very large distance from each other.[11]

(b) The hydrogen atom can exist only in states in which the total energy satisfies (Eq. 5.78).

(c) The energy difference between two different states is given by Eq. 5.79:

$$E_{n_2} - E_{n_1} = \frac{me^4}{2\hbar^2}\left(\frac{1}{n_1{}^2} - \frac{1}{n_2{}^2}\right) \tag{5.79}$$

This is the famous **Rydberg formula** for the spectral lines emitted by a hydrogen atom. We note here that the emission of light, i.e., the fact that the energy difference between two states can be converted into the energy of a photon, is not explained by *our* theory. A consistent treatment of light emission is the subject of quantum electrodynamics and goes considerably beyond the scope of this book.

(d) The ionization energy of hydrogen, i.e., the energy required to sever the electron completely, is obtained by letting $n_1 = 1$ and $n_2 = \infty$ in Eq. 5.79;

$$E_{\text{ion}} = \frac{me^4}{2\hbar^2} \tag{5.80}$$

All these findings are in splendid agreement with the experimental results, and we have every reason to consider our theory completely vindicated.[12]

Now to some of the finer points. Equation 5.73 implies that

$$n > l$$

because otherwise for

$$n = l$$

c_{n+1} and, thereby, all the following coefficients would be infinite.

[11] Note that in Eq. 3.35, we had used a different convention, making the energy of the oscillator zero for the equilibrium position. Obviously that is the preferable choice if $V(r)$ goes to infinity for large r.
[12] See also footnote 1.

Thus we have the following picture: The eigenfunctions of the hydrogen atom are

$$u_{nlm} = \frac{1}{r} \exp\left[-(me^2/n\hbar^2)r\right] \sum_0^n c_k r^k Y_{lm} \tag{5.81}$$

where the Y_{lm} are the spherical harmonics. The **principal quantum number,** n, can have any integer value:

$$n = 1, 2, 3, \ldots$$

The quantum number l can have any integer value:

$$l = 0, 1, 2, \ldots, n-1$$

and the quantum number m can have any integer value:

$$-l \leqslant m \leqslant l$$

In writing this latter statement we have followed the usual convention of writing Eq. 5.62 in the form

$$Y_{lm} = N_{lm} e^{im\varphi} P_l^{|m|}(\vartheta) \tag{5.82}$$

where m can assume any value between $m = -l$ and $m = l$, instead of writing

$$Y_{lm} = N_{lm} e^{\pm im\varphi} P_l^m(\vartheta) \tag{5.83}$$

where m can have only positive values $0 \leqslant m \leqslant l$.

The energy depends only on n, and n different values of l are possible for each value of n. For each of these values of l there are $2l + 1$ different values of m possible. To all of these different quantum numbers m and l belong different eigenfunctions (Eq. 5.81), but only one eigenvalue E_n. The hydrogen eigenvalues are thus very highly degenerate except for the lowest value E_1. The degree of degeneracy of a state with the quantum number n is obviously given by

$$\sum_0^{n-1} 2l + 1$$

It is customary to call a state with $l = 0$ an s-state, a state with $l = 1$ a p-state, and states with higher values of l, $d, f, g, h \ldots$ states.

The origin of these labels goes back to the prequantum days of spectroscopy. The letters s, p, d, f refer to the character of the spectral lines, sharp, principle, diffuse, and fine, connected with these states.

The value of the principal quantum number is usually expressed as a number preceding the letter symbol.

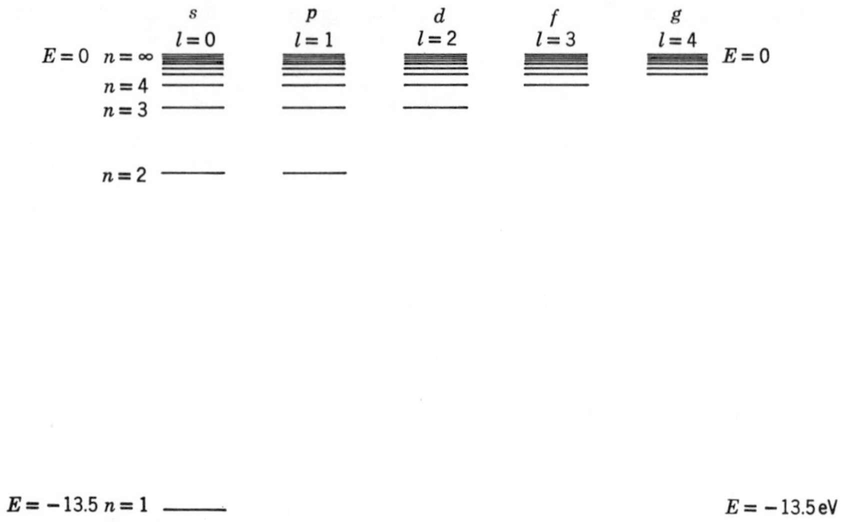

Fig. 5.3   The energy-level diagram of the hydrogen atom.

The ls state is thus the lowest energy **ground state** of the hydrogen atom. The $2s$ and $2p$ state are the states with $n = 2$, $l = 0$ and $n = 2$, $l = 1$, etc. Figure 5.3 shows the energy level diagram of the hydrogen atom.

We still owe an explanation of the physical meaning of the quantum numbers l and m. We postpone this explanation and a more detailed discussion of the hydrogen eigenfunctions until we have acquired a better understanding of some of the more formal aspects of quantum mechanics. We now content ourselves with the derivation of some of the more obvious consequences of Eq. 5.78.

5.7   HYDROGENLIKE ATOMS, THE ISOTOPE SHIFT

The term hydrogenlike atoms is used to describe atoms consisting of one positive and one negative particle. Obviously such atoms must have a Hamiltonian similar to Eq. 5.2, and all the expressions derived for the hydrogen atom must be applicable.

The He Ion

The spectrum of the singly ionized He atom[13] is described by Eq. 5.79 except that all energy eigenvalues are larger by a factor 4. In the Hamiltonian

---

[13] The neutral He atom is a three body problem and has altogether different eigenfunctions and eigenvalues.

(Eq. 5.2) we expressed the potential energy as

$$\frac{e^2}{r} = \frac{(\text{electron charge}) \cdot (\text{proton charge})}{r}$$

obviously this has to be modified to

$$\frac{e(Ze)}{r} = \frac{e^2 Z}{r} \tag{5.84}$$

if the central nucleus has a charge Z. Instead of a factor e^4, this yields a factor $e^4 Z^2$ in Eq. 5.79. In the case of the He ion ($Z = 2$) this quadruples the energy eigenvalues.

A very accurate measurement of the spectrum of the He ion shows a small deviation from this prediction. This is because we also have to replace the reduced mass:

$$m = \frac{m_e m_p}{m_e + m_p}$$

in Eq. 5.79 with

$$m' = \frac{m_e m_\alpha}{m_e + m_\alpha} \tag{5.85}$$

where m_α is the mass of the He nucleus. The deviation is

$$\frac{m'}{m} = \frac{m_\alpha}{m_e + m_\alpha} \frac{m_e + m_p}{m_p} \approx 1 + \frac{3}{4} \frac{m_e}{m_p} = 1.0004078 \tag{5.86}$$

in excellent agreement with the experimental value:

$$1.0004071$$

Muonic Atoms

Muons are particles which very much resemble electrons, except that their mass is about 200 times that of the electron. A negative muon can be attracted by a nucleus and "orbit" around it just like an electron. Since the mass of the muon is $\approx 200 m_e$, the "radius" of its orbits is 200 times smaller than that of a corresponding electron orbit. The picture of an orbiting electron or muon is of course not really correct. What we should have said is that as a result of the mass dependence of the factor $e^{-\alpha r}$ in Eq. 5.68, the eigenfunction goes faster to zero for large r if the mass m is larger. The consequences are, however, the same.

The muon "orbits" the nucleus deep inside the innermost electron shell and sees the full charge Z of the nucleus. Accordingly, muonic atoms have hydrogen-like spectra but all energy eigenvalues are higher by a factor:

$$\frac{m_\mu}{m_e} Z^2 \approx 200 Z^2$$

This leads to the emission of photons in the x-ray region.[14] Since the muon comes much closer to the nucleus than the electrons, muonic atoms are a very sensitive probe for the range of the nuclear forces. Whereas for the light elements (Eq. 5.79)—after multiplication with $(m_\mu/m_e)Z^2$—holds quite well, the deviation increases rapidly with Z. From the size of the deviation one can calculate the size of the nucleus. Some of the best measurements of nuclear diameters have been made this way.

Positronium

A positron is a positive electron. If it ever gets together with an electron, the two annihilate each other and 2 or 3 γ-rays result. Before they annihilate each other, the positron and electron sometimes form an atom called positronium, which lives for 10^{-7} or 10^{-10} sec, depending on the relative orientation of electron and positron spin. We recall that the mass, m, in Eq. 5.79 is the reduced mass:

$$m = \frac{m_e m_p}{m_e + m_p}$$

If we replace the proton mass, m_p, with the positron mass, m_e, we obtain

$$m = \frac{m_e{}^2}{2m_e} = \frac{m_e}{2} \tag{5.87}$$

Hence the energy eigenvalues of the positronium atom are $\frac{1}{2}$ of the energy eigenvalues of the hydrogen atom.

Muonium

A *positive* muon can capture an electron and form muonium. The reduced mass is in this case:

$$m = \frac{m_e m_\mu}{m_e + m_\mu} \approx m_e$$

i.e., the energy eigenvalues of a muonium atom are almost equal to those of a hydrogen atom.

The fact that the reduced mass rather than the electron mass enters into the derivation of the energy eigenvalues makes itself felt in a similar way in the spectra of nonhydrogenlike atoms. It leads to an—exceedingly small—energy difference between the spectra of different isotopes of the same element.

---

[14] We are referring here to photons emitted when the muon changes its state. Jumping electrons emit the usual spectrum of an atom with $Z - 1$.

5.8 ANGULAR MOMENTUM

We have yet to explain the physical significance of the quantum numbers m and l. The fact that they were obtained in solving the angular part of the Schrödinger equation suggests that they might have something to do with the angular momentum of the atom. To pursue this idea we translate the classical expressions for the angular momentum into the language of quantum mechanics.

The Angular Momentum Operators

In classical physics the angular momentum of a point mass with the momentum $\mathbf{p}$ is defined as

$$\mathbf{L}_{cl} = \mathbf{r} \times \mathbf{p} = \begin{vmatrix} \mathbf{i} & x & p_x \\ \mathbf{j} & y & p_y \\ \mathbf{k} & z & p_z \end{vmatrix} = \begin{pmatrix} yp_z - zp_y \\ zp_x - xp_z \\ xp_y - yp_x \end{pmatrix} \tag{5.88}$$

where $\mathbf{i}$, $\mathbf{j}$, and $\mathbf{k}$ are the unit vectors in the x, y, and z direction. The quantum mechanical equivalent is, according to the postulates of Chapter 2.1:

$$\mathbf{L} = -i\hbar \begin{pmatrix} y\dfrac{\partial}{\partial z} - z\dfrac{\partial}{\partial y} \\[2mm] z\dfrac{\partial}{\partial x} - x\dfrac{\partial}{\partial z} \\[2mm] x\dfrac{\partial}{\partial y} - y\dfrac{\partial}{\partial x} \end{pmatrix} = \begin{pmatrix} L_x \\ L_y \\ L_z \end{pmatrix} \tag{5.89}$$

We transform this to spherical polar coordinates:

$$\begin{aligned} x &= r \sin \vartheta \cos \varphi \\ y &= r \sin \vartheta \sin \varphi \\ z &= r \cos \vartheta \end{aligned} \tag{5.90}$$

From

$$r^2 = x^2 + y^2 + z^2$$

follows

$$\frac{\partial r}{\partial x} = \frac{x}{r}, \qquad \frac{\partial r}{\partial y} = \frac{y}{r}, \qquad \frac{\partial r}{\partial z} = \frac{z}{r} \tag{5.91}$$

Also, since

$$\frac{y}{x} = \tan \varphi \tag{5.92}$$

$$\frac{\partial}{\partial x}(\tan \varphi) = \frac{1}{\cos^2 \varphi} \frac{\partial \varphi}{\partial x} = -\frac{y}{x^2} \tag{5.93}$$

and

$$\frac{\partial}{\partial y}(\tan \varphi) = \frac{1}{\cos^2 \varphi} \frac{\partial \varphi}{\partial y} = \frac{1}{x} \tag{5.94}$$

therefore

$$\frac{\partial \varphi}{\partial x} = -\frac{y}{x^2} \cos^2 \varphi, \qquad \frac{\partial \varphi}{\partial y} = \frac{\cos^2 \varphi}{x} \tag{5.95}$$

We differentiate

$$z = r \cos \vartheta \tag{5.96}$$

partially with respect to x, y, and z

$$\frac{\partial z}{\partial x} = 0 = \frac{\partial r}{\partial x} \cos \vartheta - r \sin \vartheta \frac{\partial \vartheta}{\partial x} \tag{5.97}$$

or, using Eqs. 5.90 and 5.91

$$\frac{\partial \vartheta}{\partial x} = \frac{x}{r} \cdot \frac{\cos \vartheta}{r \sin \vartheta} = \frac{\cos \varphi \cos \vartheta}{r} \tag{5.98}$$

Similarly, from

$$\frac{\partial z}{\partial y} = 0 = \frac{\partial r}{\partial y} \cos \vartheta - r \sin \vartheta \frac{\partial \vartheta}{\partial y} \tag{5.99}$$

and

$$\frac{\partial z}{\partial z} = 1 = \frac{\partial r}{\partial z} \cos \vartheta - r \sin \vartheta \frac{\partial \vartheta}{\partial z} \tag{5.100}$$

follows

$$\frac{\partial \vartheta}{\partial y} = \frac{y}{r} \frac{\cos \vartheta}{r \sin \vartheta} = \frac{\sin \varphi \cos \vartheta}{r} \tag{5.101}$$

and

$$\frac{\partial \vartheta}{\partial z} = \left(\frac{z}{r} \cos \vartheta - 1\right) \frac{1}{r \sin \vartheta} = -\frac{\sin \vartheta}{r} \tag{5.102}$$

Now

$$\frac{\partial}{\partial y} = \frac{\partial r}{\partial y} \frac{\partial}{\partial r} + \frac{\partial \vartheta}{\partial y} \frac{\partial}{\partial \vartheta} + \frac{\partial \varphi}{\partial y} \frac{\partial}{\partial \varphi} \tag{5.103}$$

equivalent expressions are found for $\partial/\partial x$ and $\partial/\partial z$. Using Eqs. 5.90, 5.103, 5.91, 5.101 and 5.95 in Eq. 5.89, we obtain

$$L_x = i\hbar \left(\sin \varphi \frac{\partial}{\partial \vartheta} + \cot \vartheta \cos \varphi \frac{\partial}{\partial \varphi}\right) \tag{5.104}$$

Similarly

$$L_y = i\hbar \left(-\cos \varphi \frac{\partial}{\partial \vartheta} + \cot \vartheta \sin \varphi \frac{\partial}{\partial \varphi}\right) \tag{5.105}$$

and

$$L_z = -i\hbar \frac{\partial}{\partial \varphi} \tag{5.106}$$

L_z is immediately recognized as an operator that has the hydrogen wave functions as eigenfunctions with the eigenvalues $\pm m\hbar$ since

$$-i\hbar \frac{\partial u}{\partial \varphi} = -i\hbar\chi(r)P_l^m(\vartheta) \frac{\partial}{\partial \varphi} e^{\pm im\varphi} = \pm m\hbar u \tag{5.107}$$

in other words: **The quantum number m is a measure of the z component of the angular momentum in units of $\hbar$.**

5.9   THE OPERATOR OF THE TOTAL ANGULAR MOMENTUM

We now form the operator of the square of the total angular momentum:

$$L^2 = L_x^2 + L_y^2 + L_z^2 \tag{5.108}$$

Now

$$L_x^2 = -\hbar^2\left(\sin\varphi \frac{\partial}{\partial\vartheta} + \cot\vartheta\cos\varphi \frac{\partial}{\partial\varphi}\right)\left(\sin\varphi \frac{\partial}{\partial\vartheta} + \cot\vartheta\cos\varphi \frac{\partial}{\partial\varphi}\right)$$

$$= -\hbar^2\Bigg\{\sin^2\varphi \frac{\partial^2}{\partial\vartheta^2} + \sin\varphi\cos\varphi\cot\vartheta \frac{\partial^2}{\partial\vartheta\,\partial\varphi} - \frac{\sin\varphi\cos\varphi}{\sin^2\vartheta} \frac{\partial}{\partial\varphi}$$

$$+ \cot\vartheta\cos^2\varphi \frac{\partial}{\partial\vartheta} + \cot\vartheta\cos\varphi\sin\varphi \frac{\partial^2}{\partial\varphi\,\partial\vartheta}$$

$$+ \cot^2\vartheta\cos^2\varphi \frac{\partial^2}{\partial\varphi^2} - \cot^2\vartheta\cos\varphi\sin\varphi \frac{\partial}{\partial\varphi}\Bigg\} \tag{5.109}$$

Also

$$L_y^2 = -\hbar^2\Bigg\{\cos^2\varphi \frac{\partial^2}{\partial\vartheta^2} - \cos\varphi\cot\vartheta\sin\varphi \frac{\partial^2}{\partial\varphi\,\partial\vartheta}$$

$$+ \cot\vartheta\sin^2\varphi \frac{\partial}{\partial\vartheta} - \cot\vartheta\sin\varphi\cos\varphi \frac{\partial^2}{\partial\varphi\,\partial\vartheta}$$

$$+ \frac{\cos\varphi\sin\varphi}{\sin^2\vartheta} \frac{\partial}{\partial\varphi} + \cot^2\vartheta\sin^2\varphi \frac{\partial^2}{\partial\varphi^2}$$

$$+ \cot^2\vartheta\cos\varphi\sin\varphi \frac{\partial}{\partial\varphi}\Bigg\} \tag{5.110}$$

Finally

$$L_z^2 = -\hbar^2 \frac{\partial^2}{\partial\varphi^2} \tag{5.111}$$

Adding Eqs. 5.109, 5.110, and 5.111 gives

$$L^2 = -\hbar^2 \left\{ \frac{\partial^2}{\partial \vartheta^2} + \cot \vartheta \frac{\partial}{\partial \vartheta} + \frac{1}{\sin^2 \vartheta} \frac{\partial^2}{\partial \varphi^2} \right\} \qquad (5.112)$$

This can also be written

$$L^2 = -\hbar^2 \left\{ \frac{1}{\sin \vartheta} \frac{\partial}{\partial \vartheta} \left(\sin \vartheta \frac{\partial}{\partial \vartheta} \right) + \frac{1}{\sin^2 \varphi} \frac{\partial^2}{\partial \varphi^2} \right\} \qquad (5.113)$$

We compare this with the angular part of the Schrödinger equation:

$$\frac{1}{\sin \vartheta} \cdot \frac{\partial}{\partial \vartheta} \left(\sin \vartheta \frac{\partial Y}{\partial \vartheta} \right) + \frac{1}{\sin^2 \vartheta} \cdot \frac{\partial^2 Y}{\partial \varphi^2} = -l(l + 1)Y \qquad (5.26)$$

The conclusion is obvious, the angular part of the hydrogen wave functions $Y_{lm}(\vartheta \varphi)$ and hence the entire wave functions, $u(\mathbf{r})$, are eigenfunctions of the operator L^2

$$L^2 u(\mathbf{r}) = \chi(r) L^2 Y_{lm}(\vartheta, \varphi) = \hbar^2 l(l + 1) u(\mathbf{r}) \qquad (5.114)$$

with the eigenvalues

$$l(l + 1)\hbar^2$$

We conclude: **The quantum number l is a measure of the total angular momentum in units of $\hbar$.**

Considering the remarks made on p. 76, we can now more fully appreciate the implications of the separability of Eq. 5.22. The Hamiltonian can be written as the sum of a radial part and the angular momentum operator Eq. 5.113. This connects the separability with the conservation of angular momentum.[15] The separation constant is quantized in this case so that the angular and the radial part are not fully independent but are connected by the condition that $A = l(l + 1)$ in both Eq. 5.25 and Eq. 5.26.

5.10   THE EIGENFUNCTIONS OF HYDROGEN, ATOMIC UNITS

Now that we understand the meaning of the various quantum numbers of the hydrogen atom, we return to the discussion of its eigenfunctions. Table 5.1 lists the complete hydrogen eigenfunctions $u(r, \vartheta, \varphi)$ for the four lowest values of n. The normalization constants have been determined such that $\int u^* u \, d\tau = 1$. In the absence of an interaction with some external force any

---

[15] We shall explore this connection in more detail in Chapter 6.

Table 5.1

State	n	l	m	u	
1s	1	0	0	$A_n e^{-x}$	$x = \dfrac{rme^2}{n\hbar^2}$
2s	2	0	0	$A_n e^{-x}(1-x)$	
2p	2	1	0	$A_n e^{-x} x \cos\vartheta$	$A_n = \dfrac{1}{\sqrt{\pi}} \left(\dfrac{me^2}{n\hbar^2}\right)^{3\!/\!2}$
2p	2	1	± 1	$A_n \dfrac{e^{-x}}{\sqrt{2}} x \sin\vartheta e^{\pm i\varphi}$	
3s	3	0	0	$A_n e^{-x}\left(1 - 2x + \dfrac{2x^2}{3}\right)$	
3p	3	1	0	$A_n e^{-x}\sqrt{\dfrac{2}{3}}\, x(2-x)\cos\vartheta$	
3p	3	1	± 1	$A_n e^{-x}\dfrac{1}{\sqrt{3}} x(2-x)\sin\vartheta e^{\pm i\varphi}$	
3d	3	2	0	$A_n e^{-x}\dfrac{1}{3\sqrt{2}} x^2(3\cos^2\vartheta - 1)$	
3d	3	2	± 1	$A_n e^{-x}\dfrac{x^2}{\sqrt{3}}\sin\vartheta\cos\vartheta e^{\pm i\varphi}$	
3d	3	2	± 2	$A_n e^{-x}\dfrac{1}{2\sqrt{3}} x^2\sin^2\vartheta e^{\pm 2i\varphi}$	
4s	4	0	0	$A_n e^{-x}\left(1 - 3x + 2x^2 - \dfrac{x^3}{3}\right)$	
4p	4	1	0	$A_n e^{-x}\sqrt{5}x\left(1 - x + \dfrac{x^2}{5}\right)\cos\vartheta$	
4p	4	1	± 1	$A_n e^{-x}\sqrt{\dfrac{5}{2}}x\left(1 - x + \dfrac{x^2}{5}\right)\sin\vartheta e^{\pm i\varphi}$	
4d	4	2	0	$A_n e^{-x}\dfrac{1}{2} x^2\left(1 - \dfrac{x}{3}\right)(3\cos^2\vartheta - 1)$	
4d	4	2	± 1	$A_n e^{-x}\sqrt{\dfrac{3}{2}} x^2\left(1 - \dfrac{x}{3}\right)\sin\vartheta\cos\vartheta e^{\pm i\varphi}$	
4d	4	2	± 2	$A_n e^{-x}\sqrt{\dfrac{3}{8}} x^2\left(1 - \dfrac{x}{3}\right)\sin^2\vartheta e^{\pm 2i\varphi}$	
4f	4	3	0	$A_n e^{-x}\dfrac{1}{6\sqrt{5}} x^3\cos\vartheta\,(5\cos^2\vartheta - 3)$	
4f	4	3	± 1	$A_n e^{-x}\dfrac{1}{6}\sqrt{\dfrac{3}{20}} x^3\sin\vartheta\,(5\cos^2\vartheta - 1)e^{\pm i\varphi}$	
4f	4	3	± 2	$A_n e^{-x}\dfrac{\sqrt{3}x^3}{6\sqrt{2}}\sin^2\vartheta\cos\vartheta e^{\pm 2i\varphi}$	
4f	4	3	± 3	$A_n e^{-x}\tfrac{1}{12}x^3\sin^3\vartheta e^{\pm 3i\varphi}$	

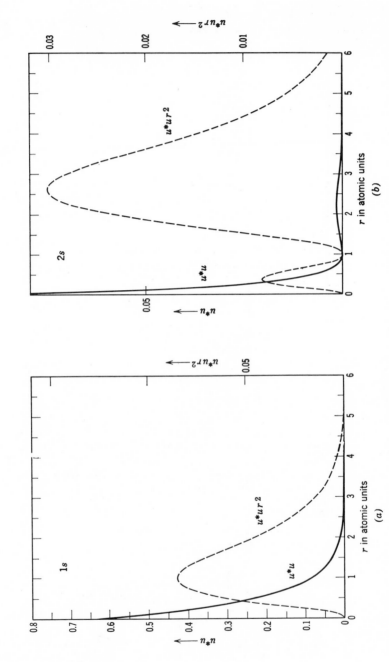

Fig. 5.4 The probability $u^*(r)u(r)r^2$ and the probability density $u^*(r)u(r)$ of the electron in a hydrogen atom as a function of the radius.

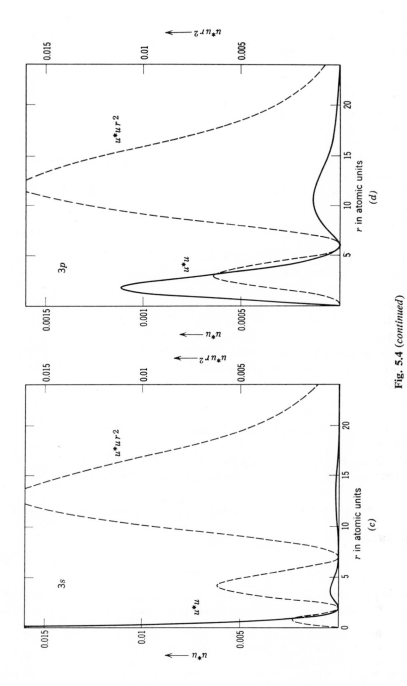

Fig. 5.4 (continued)

coordinate system should be as good as the next one. We would, therefore, expect that the electron density u^*u is independent of the angles ϑ and φ. For φ this is easy to see since

$$u^*u = \chi^2_{(r)}[P_l^m(\vartheta)]^2 e^{im\varphi} \cdot e^{-im\varphi} \tag{5.115}$$

is indeed independent of φ. It is, however, not usually[16] independent of ϑ, nor need it be. In any state with $n > 1$, there are several states with different l and m that have the same energy, i.e., are degenerate. As long as we do not, by some interaction, remove this degeneracy, the atom is not in any one of these states in particular but in a mixture of all of them. The reader should convince himself that the probability density of, for instance, the $3d$ state is, indeed, independent of ϑ if we form:[17]

$$u^*u = u^*(3, 2, 0)u(3, 2, 0) + 2u^*(3, 2, 1)u(3, 2, 1) + 2u^*(3, 2, 2)u(3, 2, 2) \tag{5.116}$$

where $u(3, 2, 1)$ stands for $u(n = 3, l = 2, m = 1)$. We shall see later (Chapter 10.2), that it is also possible (for instance, with the help of an external magnetic field) to make sure that an atom is in a state with a well-defined value of l and m. (It is this case that Figure 5.5c refers to.) Figure 5.4 gives the probability density u^*u averaged over all angles ϑ as a function of r for various values of the quantum numbers n and l. Figure 5.5 attempts to give an impression of the spatial distribution of the electron probability density for various values of the quantum numbers. The probability of finding the electron in a volume element is given by

$$dP = u^*u \, d\tau = u^*u r^2 \, dr \sin \vartheta \, d\vartheta \, d\varphi \tag{5.117}$$

This probability is also plotted in Figure 5.4 as a function of r for various values of n and l.

It is interesting to calculate the radius at which the probability[18] of finding the electron, is greatest. We do this for the case of the $1s$ state:

$$0 = \frac{d}{dr}(u^*u \, d\tau) = \frac{d}{dr}\left[\exp\left(-\frac{2me^2r}{\hbar^2}\right)r^2 \, dr \sin \vartheta \, d\vartheta \, d\varphi\right] \tag{5.118}$$

This yields

$$r = \frac{\hbar^2}{me^2} \tag{5.119}$$

This happens to be the radius of the first electron orbit in Bohr's *old quantum theory*.[19] The **Bohr radius** $\hbar^2/m_e e^2$ is frequently used as the unit of length in

---

[16] Only for s-states.
[17] Where do the factors 2 come from?
[18] Not the probability density!
[19] Neglecting the small difference between the electron mass m_e and the reduced mass m.

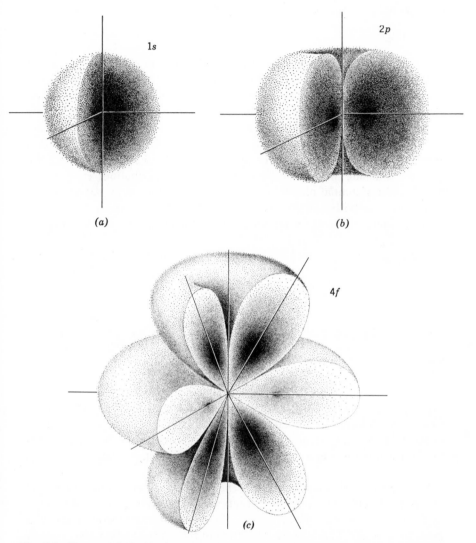

Fig. 5.5 The spatial distribution of the electron probability density u^*u in a hydrogen atom. The scale has been changed to make all three figures appear the same size. Also the shading has been normalized so that the area of highest electron density appears black. The reader may verify just how much the absolute size and density vary from state to state.

atomic physics and is the basis of the so-called **atomic units.** In atomic units the energy is measured in multiples of the ionization energy of hydrogen $m_e e^4/2\hbar^2$.[19] The use of atomic units amounts to letting $\hbar^2 = 1$, $e^2 = 2$, and $m_e = \frac{1}{2}$ in all equations. Expressed in atomic units the Schrödinger equation

Eq. 5.20 becomes for instance:[19]

$$-\nabla^2 u - \frac{2u}{r} = Eu \tag{5.120}$$

The abscissa in Figure 5.4 is divided in units of $\hbar^2/m_e e^2$. We had inquired earlier (Chapter 5.6) into the degree of degeneracy of the hydrogen eigenfunctions. A look at Table 5.1 shows a marked difference between the m and the l degeneracy. The m degeneracy is simply due to the symmetry of the eigenfunctions just as in the classical example of Chapter 4.2. The reason for the l degeneracy is less obvious. The probability distributions for the 3s and the 3p state for instance are completely different (see Figure 5.4c, d). We can only conclude that the shape of the electron distributions is such that for the given potential of the form $V = e^2/r$ the total energy *just happens* to be the same for both states. To demonstrate how this can occur, we calculate the expectation value of the potential energy for the 2s and the 2p state.

$$\langle V \rangle_{2p} = \frac{N^2}{\pi} \int e^{-2ar} a^2 r^2 \cos^2 \vartheta r^2 \, dr \sin \vartheta \, d\vartheta \, d\varphi$$

$$= \frac{N^2}{\pi} a^2 2\pi \frac{2}{3} \int e^{-2ar} r^4 \, dr = \frac{N^2}{a^3} \tag{5.121}$$

$$\langle V \rangle_{2s} = \frac{N^2}{\pi} \int e^{-2ar}(1 - ar)^2 r^2 \, dr \sin \vartheta \, d\vartheta \, d\varphi = \frac{N^2}{a^3} \tag{5.122}$$

where

$$a = \frac{me^2}{n\hbar^2}$$

The weighted average value of the potential energy is thus the same for the two states. This must mean that the small hump in the probability distribution of the 2s state in the region of strong field makes up for the fact that most of the 2s distribution is in a weaker field than the 2p distribution.

Very accurate measurements of the energy levels reveal that the degeneracy of the states is lifted by small perturbations. These perturbations affect some of the linear combinations of the eigenfunctions of a degenerate eigenvalue more than others. This leads to a splitting of the corresponding energy levels. We shall discuss some of these effects in detail in the later chapters.

PROBLEMS

5.1 Normalize the first three Legendre polynomials using the condition (Eq. 5.48).

5.2 Show explicitly that the first three Legendre polynomials are orthogonal over the interval $-1 \leqslant \xi \leqslant 1$. Are they also orthogonal over the interval $0 \leqslant \xi \leqslant 1$?

5.3 How can we calculate the ionization energy of the hydrogen atom from spectroscopic measurements?

5.4 The spectra of muonic atoms allow us to determine the size of the nucleus. How? Explain qualitatively.

5.5 Is a positronium atom larger or smaller than a hydrogen atom? Explain.

5.6 You are given the task of finding out whether a certain element is a mixture of several isotopes. For evidence, would you turn to atomic or molecular spectra? Why?

5.7 Verify, by application of the appropriate operators, that the quantum numbers of the $2p$ state of the hydrogen atom are indeed $l = 1$, and $m = 0$, ± 1.

5.8 A baseball is pitched so that it spins at 300 rpm. Making sensible assumptions about the structural parameters of the ball (mass, composition, etc.) determine its angular momentum in units of $\hbar$. How much angular momentum is this per atom?

5.9 When an atom of a heavy element is bombarded with electrons of sufficient energy, it sometimes happens that one of the two $1s$ electrons is knocked out of the atom. The resulting hole is filled by one of the $2p$ electrons. The x-rays emitted in this process are known as the K x-rays. Derive a formula for the wavelength of the K x-rays as a function of the atomic number Z.

5.10 Calculate the expectation value of the kinetic and the potential energy of the hydrogen ground state.

5.11 What is the parity of the first three eigenfunctions in Table 5.1?

5.12 Verify that the $1s$ eigenfunction in Table 5.1 is normalized and orthogonal to the $2s$ eigenfunction.

5.13 Show that the three $3p$ eigenfunctions in Table 5.1 are orthogonal to each other.

5.14 Verify that the $4f$ eigenfunctions in Table 5.1 are normalized.

5.15 What is the degree of degeneracy of the hydrogen eigenfunctions in the states with $n = 1, 2, 3$?

5.16 Show that a hydrogen atom in the $3d$ state is spherically symmetric as long as no external perturbation distinguishes a particular coordinate system.

5.17 In Bohr's old quantum theory an electron was thought to move in a circular orbit around the nucleus. Studying Table 5.1, you will find certain eigenfunctions for which the picture of an orbiting electron is particularly appropriate. What are their quantum numbers?

5.18 (a) Derive an asymptotic expression for the radius at which the electron is most likely to be found in states with $l = n - 1$, for $n \gg 1$. (*Hint*. Study Table 5.1.) Calculate the *classical* expression for the frequency of revolution of an electron moving in a classical orbit with this radius.
(b) Derive an asymptotic expression for the frequency difference

$$\frac{(E_n - E_{n+1})}{\hbar} = \Delta\omega$$

(c) Compare the results of (a) and (b) above. Comment!

5.19   Show that whenever a Hamiltonian[20] $H(x_1, \ldots, x_n)$ can be written as a sum of two parts $H_1(x_1, \ldots, x_k) + H_2(x_{k+1}, \ldots, x_n)$ depending on different variables, the eigenvalue equation

$$H(x_1, \ldots, x_n)\psi(x_1, \ldots, x_n) = E\psi(x_1, \ldots, x_n)$$

can be separated by substituting

$$\psi_1(x_1, \ldots, x_k)\psi_2(x_{k+1}, \ldots, x_n) = \psi(x_1, \ldots, x_n).$$

How are the eigenvalues of the operators $H_1(x_1, \ldots, x_k)$, $H_2(x_{k+1}, \ldots, x_n)$, and $H(x_1, \ldots, x_n)$ related?

SOLUTIONS

5.6   The mass M of the nucleus enters the energy levels of an atom only through the reduced mass (Eq. 5.8):

$$m = \frac{m_e M}{m_e + M}$$

Since $m_e \ll M$ the small difference in M between different isotopes has a negligible influence on m. The atomic spectra of the various isotopes of an atom differ, thus, very little. A molecule in its vibrational modes approaches an harmonic oscillator. The energy eigenvalues of the harmonic oscillator, according to Eqs. 3.64, 3.43, and 3.41,

$$E_n = \frac{(2n + 1)\hbar\omega}{2} = \frac{(2n + 1)\hbar}{2}\sqrt{\frac{C}{m}}$$

are inversely proportional to the square root of the mass m of the vibrating particle.

Assume that ΔM is the mass difference between two isotopes and that $\Delta M \ll M$. In this case the ratio of the molecular energy values for the two isotopes will be

$$R = \sqrt{\frac{M + \Delta M}{M}} \approx 1 + \frac{1}{2} \cdot \frac{\Delta M}{M}$$

For the same ratio we get in the case of an atomic spectrum

$$\left[\frac{m_e(M + \Delta M)}{m_e + M + \Delta M}\right]\left(\frac{m_e + M}{m_e \cdot M}\right) \approx 1 + \frac{m_e}{M} \cdot \frac{\Delta M}{M}$$

5.13   In the product $u_{2,1}^* \cdot u_{2,-1}$ the φ dependent part becomes

$$(e^{i\varphi})^*(e^{-i\varphi}) = e^{-2i\varphi}$$

[20] Or other operator.

In the products $u_{2,0}^* \cdot u_{2,1}$ and $u_{2,0}^* \cdot u_{2,-1}$ the φ dependent part becomes

$$e^{i\varphi} \quad \text{or} \quad e^{-i\varphi}$$

In each case the integral from $\varphi = 0$ to $\varphi = 2\pi$ vanishes, i.e., the eigenfunctions are orthogonal.

5.16 The φ-dependence is obviously eliminated in the products u^*u. The only angular dependence that could exist would be a ϑ-dependence. Dropping the identical x-dependent parts, we get for the state with $m = 0$

$$u_1^* u_1 = \frac{1}{9 \cdot 2} (9 \cos^4 \vartheta - 6 \cos^2 \vartheta + 1)$$

For the states with $m = \pm 1$

$$u_2^* u_2 = \tfrac{1}{3} \sin^2 \vartheta \cos^2 \vartheta$$

and for the states with $m = \pm 2$

$$u_3^* u_3 = \frac{1}{4 \cdot 3} \sin^4 \vartheta$$

These probability densities are quite obviously ϑ-dependent. On the other hand all five states, $m = 0$, $m = \pm 1$, and $m = \pm 2$ are degenerate in the absence of any perturbation that favors a particular direction. The atom will, therefore, be in a state that is the sum of equal parts of all five states. Thus the total probability density becomes

$$u^*u = \frac{1}{9 \cdot 2} (9 \cos^4 \vartheta - 6 \cos^2 \vartheta + 1) + \frac{2}{3} \sin^2 \vartheta \cos^2 \vartheta + \frac{2}{4 \cdot 3} \sin^4 \vartheta$$

Substituting

$$\sin^2 \vartheta = 1 - \cos^2 \vartheta$$

we get

$$\tfrac{1}{2} \cos^4 \vartheta - \tfrac{1}{3} \cos^2 \vartheta + \tfrac{1}{18} + \tfrac{2}{3} \cos^2 \vartheta - \tfrac{2}{3} \cos^4 \vartheta$$
$$+ \tfrac{1}{6} + \tfrac{1}{6} \cos^4 \vartheta - \tfrac{1}{3} \cos^2 \vartheta = u^*u$$

Adding this up we find that the ϑ-dependence cancels and we obtain

$$u^*u = \tfrac{2}{9}$$

i.e., the electron distribution in the $3d$ state is isotropic.

6

MORE THEOREMS

In the preceding chapters we have tried to establish a firm connection between the more or less intuitively introduced postulates of Chapters 2 and 4 and the results of experimental physics. In doing this, we have looked neither left nor right. Lest the reader get the impression that quantum mechanics is a collection of postulates and mathematical tricks, we shall now fill in the gaps and try to show quantum mechanics as it actually is: a physical theory of great mathematical beauty, fully as elegant as classical mechanics in its Lagrangian or Hamiltonian form. We shall also discuss some mathematical techniques that will be of help in the solution of the more intricate problems in the following chapters.

6.1 ORTHONORMAL FUNCTIONS, COMPLETE SETS

Frequently we have encountered integrals of the type

$$\int \psi^* \varphi \, d\tau = c \qquad (6.1)$$

in the foregoing chapters. Sometimes the function φ satisfied $\varphi = \psi$, in which case we had normalized ψ such that $c = 1$. Sometimes the nature of φ was such that $c = 0$, in which case we had called ψ and φ orthogonal. All of this is vaguely reminiscent of the scalar product (dot product) of vector algebra. For instance

$$\int \psi^* \varphi \, d\tau = c \qquad (6.2)$$

is a constant, although ψ and φ are functions of $\mathbf{r}$, whereas

$$(\mathbf{a} \cdot \mathbf{b}) = \sum_{i=1}^{3} a_i b_i = c \qquad (6.3)$$

is a scalar, although $\mathbf{a}$ and $\mathbf{b}$ are vectors. Actually this similarity is more than skin deep and is explored in the mathematical theory of Hilbert spaces. We shall just use it as a handy excuse to apply the convenient notation and terminology of vector algebra to our problems.

Definition.   The **scalar product** c of two functions $\psi(r)$ and $\varphi(r)$ is defined as

$$\int \psi^* \varphi \, d\tau = c \tag{6.2}$$

and we shall henceforth use the abbreviation[1]

$$\int \psi^* \varphi \, d\tau = (\psi, \varphi) = c \tag{6.4}$$

for it. **Note that this notation implies that the first of the two factors is taken as the complex conjugate.** We can extend the notation in the following way to include operators

$$\int \psi^* Q\varphi \, d\tau = (\psi, Q\varphi) \tag{6.5}$$

$$\int (Q\psi)^* \varphi \, d\tau = (Q\psi, \varphi) \tag{6.6}$$

Thus the definition of a hermitian operator[2] becomes

$$(\psi, H\varphi) = (H\psi, \varphi) \tag{6.7}$$

Definition.   A set of functions $\psi_1, \psi_2, \ldots, \psi_n$ is said to be **linearly independent** if there exist no constants $a_1, a_2, \ldots, a_n$, different from zero, so that

$$\sum_1^n a_k \psi_k(\mathbf{r}) = 0 \tag{6.8}$$

is satisfied for all choices of $\mathbf{r}$. Obviously this means that none of the functions ψ_k can be expressed as a linear combination of all the others.

Definition.   A set of functions $\psi_1, \psi_2, \ldots, \psi_n$ that satisfy

$$(\psi_i, \psi_k) = \delta_{ik} \begin{cases} = 1 \; if \; i = k \\ = 0 \; if \; i \neq k \end{cases} \tag{6.9}$$

is said to be **orthonormal.**

Let $\psi_1, \psi_2, \ldots, \psi_n$ be a set of linearly independent functions. Let

$$\varphi = \sum_1^n a_k \psi_k \tag{6.10}$$

---

[1] The notation implies that the integral is taken over all space.
[2] See Appendix A.1.

be a linear combination of the ψ_k. If the ψ_k form an orthonormal set, it is especially easy to determine the coefficients a_k in Eq. 6.10. To this end we multiply both sides with, for example, $\psi_i{}^*$ and integrate:

$$(\psi_i, \varphi) = \sum_1^n a_k(\psi_i, \psi_k) \tag{6.11}$$

Because of the orthogonality of the ψ_k all terms of the sum except the one with $\psi_k = \psi_i$ vanish. Because of the normalization of the ψ_k

$$(\psi_i, \psi_i) = 1 \tag{6.12}$$

hence

$$a_i = (\psi_i, \varphi) \tag{6.13}$$

It is obvious that the larger the set of the ψ_i is, the more linearly independent functions φ we can construct with it in the manner of Eq. 6.10. It is interesting to raise the question whether with an infinite set of orthonormal functions ψ_k, we can express any arbitrary function $\varphi(\mathbf{r})$ as

$$\varphi(\mathbf{r}) = \sum_1^\infty a_k \psi_k \tag{6.14}$$

Obviously we can do this if for any function $\varphi(\mathbf{r})$ we can make

$$\lim_{n \to \infty} \left(\varphi(\mathbf{r}) - \sum_1^n a_k \psi_k \right) \equiv 0 \tag{6.15}$$

through a suitable choice[3] of the a_k. Equation 6.15 is not very suitable as a practical criterion. If we want a workable expression, we have to settle for less:

$$\lim_{n \to \infty} \left(\int |\varphi - \sum a_k \psi_k|^2 \, d\tau \right) = 0 \tag{6.16}$$

where, in order to carry out the integration, we have to assume $\varphi(r)$ to be piecewise continuous. Whereas Eq. 6.15 would make $\varphi(\mathbf{r}) = \sum a_k \psi_k$ *everywhere*, the slightly less stringent condition, Eq. 6.16 allows for the occurrence of Gibbs' phenomenon[4] in the vicinity of a discontinuity in $\varphi(\mathbf{r})$. Now we form

$$\int \left| \varphi - \sum_1^n a_k \psi_k \right|^2 d\tau = \int \left(\varphi^* - \sum_1^n a_k{}^* \psi_k{}^* \right) \left(\varphi - \sum_1^n a_k \psi_k \right) d\tau = \epsilon_n \tag{6.17}$$

---

[3] The reader who is still not familiar with the contents of Appendixes A.1, A.2, and A.3 should study them *now*.

[4] See Appendix A.2.

Using Eqs. 6.9 and 6.13, this becomes

$$\epsilon_n = (\varphi, \varphi) - \sum_1^n a_k a_k^* - \sum_1^n a_k^* a_k + \sum_1^n a_k^* a_k \tag{6.18}$$

or

$$\epsilon_n = (\varphi, \varphi) - \sum a_k^* a_k \tag{6.19}$$

hence Eq. 6.16 is satisfied if

$$\lim_{n \to \infty} \epsilon_n = 0 \tag{6.20}$$

or

$$(\varphi, \varphi) = \sum_1^\infty a_k^* a_k \tag{6.21}$$

This leads to the next definition.

Definition.   An orthonormal set of functions ψ_k, $k = 1, 2, 3, \ldots$ is said to be **complete** in a volume V if for any piecewise continuous function $\varphi(r)$

$$(\varphi, \varphi) = \sum_1^\infty a_k^* a_k \tag{6.22}$$

where the integral is taken over the volume V and the a_k are defined by

$$a_k = (\psi_k, \varphi) \tag{6.23}$$

Keeping in mind the subtle difference between Eqs. 6.15 and 6.16, we brazenly interpret the completeness relation (Eq. 6.21) to mean that any function (at least of the kind we have to deal with) can be expanded in the form:

$$\varphi = \sum_1^\infty a_k \psi_k \tag{6.14}$$

where the a_k are given by Eq. 6.13.

It is obvious that despite its mathematical elegance the completeness relation is almost useless as a practical criterion. How are we to decide whether Eq. 6.19 is valid for *any* function φ? We leave this thorny problem to the mathematicians and merely convince ourselves below that the eigenfunctions of operators of physical significance form orthonormal sets, emphasizing, however, that the orthonormality of a set of functions does not necessarily imply its completeness.[5] The proof that any set of functions is

---

[5] The statement that an infinite orthonormal set is not necessarily complete is very easy to prove

$$\cos(\omega t), \cos(2\omega t), \ldots, \cos((k-1)\omega t), \cos((k+1)\omega t), \cos((k+2)\omega t), \text{etc.}$$

is certainly infinite and orthogonal. It can also be normalized (see Appendix A.2.7); yet, it is not complete since it cannot be used to expand the function $\cos(k\omega t)$. Similarly, $\cos(k\omega t)$, $k = 0, 1, 2, 3, \ldots$ is not a complete set, since it is obvious from changing t into $-t$ that

$$\sin(n\omega t) = \sum_0^\infty a_k \cos(k\omega t)$$

cannot be satisfied for $n \neq 0$.

complete is usually rather difficult (not surprising if we consider how sweeping the statement of completeness is). However, the eigenfunctions of the operators we are dealing with in quantum mechanics have, generally, been shown to form complete sets, at least with regard to the kind of wave functions we are interested in, and we shall frequently expand functions in terms of such sets without further proof of their completeness.

By now it should be apparent to the reader that the expansion of a function in terms of a complete set of other functions is merely a generalization of the Fourier expansion discussed in Appendix A.2.

At first glance it may seem difficult to lend substance to the above statement about the eigenfunctions of *operators of physical significance*. Is there any one mathematical characteristic that is common to all operators that have a physical meaning? Fortunately, there is.

Let Q be an operator describing some measurable physical quantity, for example the momentum, of a system. The expectation value

$$\langle Q \rangle = (\psi_i, Q\psi_i) \tag{6.24}$$

is then the result that we can expect as the average of many measurements of that quantity (Chapter 4.1). If ψ_i is an eigenfunction of the operator Q, Eq. 6.24 yields, of course, $\langle Q \rangle = q_i$ where q_i is the eigenvalue of ψ_i. This shows that the eigenvalues of operator can be measurable quantities. An operator of *physical significance* is obviously one whose eigenvalues *are* measurable quantities. Measurable quantities, whatever they may be, are certainly real. Operators of physical significance must, therefore, have one thing in common: their eigenvalues must be real. Once a physical problem has been properly phrased mathematically, it is more likely than not that the answer to it has been available for 100 years or so. In our case the answer is given by the following theorem.

THEOREM

The eigenvalues of a hermitian operator Q are real.

Proof

$$(\psi, Q\psi) = (\psi, q\psi) = q(\psi, \psi) \tag{6.25}$$

also

$$(Q\psi, \psi) = (q\psi, \psi) = q^*(\psi, \psi) \tag{6.26}$$

Now, since Q is hermitian,

$$(\psi, Q\psi) = (Q\psi, \psi) \quad \text{and hence} \quad q = q^*$$

The above proof can also be read from right to left yielding the statement: An operator, whose eigenvalues are real, is hermitian.

This leaves us with the comfortable knowledge that the theorems that we shall prove below for hermitian operators cover all the cases of physical interest.

For good measure we prove that two of the most important operators are indeed hermitian.

THEOREM

The Hamiltonian operator

$$H = -\frac{\hbar^2}{2m}\nabla^2 + V(\mathbf{r}) \tag{6.27}$$

is hermitian.

Proof. $V(r)$ is a real multiplicative factor and thereby hermitian. The proof of the hermiticity of the Laplace operator ∇^2 is given in Appendix A.1.

THEOREM

The momentum operator $-i\hbar\nabla$ *is hermitian.*

Proof

$$\frac{\partial}{\partial x}(\psi^*\psi) = \psi^*\frac{\partial\psi}{\partial x} + \psi\frac{\partial\psi^*}{\partial x} \tag{6.28}$$

Integration of Eq. 6.28 yields

$$\int\frac{\partial}{\partial x}(\psi^*\psi)\,d\tau = \int\psi^*\frac{\partial\psi}{\partial x}\,d\tau + \int\psi\frac{\partial\psi^*}{\partial x}\,d\tau \tag{6.29}$$

The integral on the left side vanishes because $\psi^*\psi$ vanishes for large values of $|x|$ if the particle is confined to some finite region:

$$\int\frac{\partial}{\partial x}(\psi^*\psi)\,dx\,dy\,dz = \int|\psi^*\psi|_{x=-\infty}^{x=\infty}dy\,dz = 0 \tag{6.30}$$

Hence

$$\int\psi^*\frac{\partial\psi}{\partial x}\,d\tau = -\int\psi\frac{\partial\psi^*}{\partial x}\,d\tau \tag{6.31}$$

or, since corresponding expressions are obtained for the other coordinates,

$$(\psi, \nabla\psi) = -(\nabla\psi, \psi) \tag{6.32}$$

hence[6]

$$(\psi, i\hbar\,\nabla\psi) = (i\hbar\,\nabla\psi, \psi) \qquad \text{q.e.d.} \tag{6.33}$$

---

[6] Remember that the complex conjugate of the first factor is to be taken, i.e., $(i\hbar\,\nabla\psi, \psi) = -\int i\hbar\,\nabla\psi^*\psi\,d\tau$.

From the hermiticity of the momentum operator follows immediately that the operators L_x, L_y and L_z are hermitian.

Proof. According to Eq. 5.89,

$$L_x = -i\hbar y \frac{\partial}{\partial z} + i\hbar z \frac{\partial}{\partial y}$$

Since the operators $p_z = -i\hbar(\partial/\partial z)$ and $p_y = -i\hbar(\partial/\partial y)$ are hermitian, and since y and z are real factors and, thereby, hermitian, it follows that L_x is hermitian. The proof for L_y and L_z follows from a cyclic permutation of the variables. (A cyclic permutation replaces x with y, y with z, and z with x.) Since the angular momentum operators can be derived from each other through a cyclic permutation any expression derived for one of them can be transformed into an expression for the others by means of a cyclic permutation. The hermiticity of L^2 follows immediately from the hermiticity of L (see Problem 6.2).

THEOREM

The eigenfunctions ψ_1, ψ_2, ψ_3, ... of a hermitian operator Q, belonging to different eigenvalues $q_1, q_2, q_3, \ldots$ are orthogonal over Q's region of hermiticity.[7]

Proof. The proof is accomplished by showing the above statement to be true for *any* two eigenfunctions ψ_i and ψ_k.

$$(Q\psi_i, \psi_k) = q_i{}^*(\psi_i, \psi_k) = q_i(\psi_i, \psi_k) \tag{6.34}$$

also

$$(\psi_i, Q\psi_k) = q_k(\psi_i, \psi_k)$$

Because of the hermiticity of Q:

$$(\psi_i, Q\psi_k) = (Q\psi_i, \psi_k) \quad \text{or} \quad (q_i - q_k)(\psi_i, \psi_k) = 0 \tag{6.35}$$

Since by assumption $q_i \neq q_k$, it follows that

$$(\psi_i, \psi_k) = 0 \quad \text{for} \quad i \neq k \quad \text{q.e.d.}$$

We still have to cover ourselves for the eventuality that the eigenvalues are degenerate. Let ψ_i and ψ_k be two normalized linearly independent eigenfunctions of the hermitian operator Q, both belonging to the same eigenvalue q, i.e.,

$$Q\psi_i = q\psi_i, \quad Q\psi_k = q\psi_k \tag{6.36}$$

---

[7] Sometimes the condition $\int_V \psi^*(Q\varphi)\, d\tau = \int_V (Q\psi)^*\varphi\, d\tau$ is satisfied only in a certain volume V and not elsewhere. In this case V is referred to as Q's region of hermiticity.

Since our above proof breaks down ψ_i and ψ_k need not be orthogonal even if Q is hermitian. We form

$$Q(\alpha\psi_i + \beta\psi_k) = q(\alpha\psi_i + \beta\psi_k) = q\psi_{k'} \qquad (6.37)$$

where

$$\psi_{k'} = \alpha\psi_i + \beta\psi_k \qquad (6.38)$$

Thus, $\psi_{k'}$ is a new eigenfunction to the same eigenvalue q. The constants α and β are arbitrary, and we determine them by requiring that

$$(\psi_i, \psi_{k'}) = 0 \qquad (6.39)$$

and

$$(\psi_{k'}, \psi_{k'}) = 1 \qquad (6.40)$$

Because of the normalization of ψ_i and ψ_k this yields

$$\alpha = -\beta(\psi_i, \psi_k) \qquad (6.41)$$

and

$$\alpha^2 + \alpha\beta[(\psi_i, \psi_k) + (\psi_k, \psi_i)] + \beta^2 = 1 \qquad (6.42)$$

If we solve Eqs. 6.41 and 6.42 for α and β and substitute into Eq. 6.68, we obtain

$$\psi_{k'} = \frac{\psi_k - (\psi_i, \psi_k)\psi_i}{\sqrt{1 - |(\psi_i, \psi_k)|^2}} \qquad (6.43)$$

This is an eigenfunction of Q with the eigenvalue q, it is orthogonal to ψ_i, and it is normalized. We can, thus, amend the above theorem as follows: The eigenfunctions of a hermitian operator belonging to nondegenerate eigenvalues are orthogonal. From the eigenfunctions that belong to degenerate eigenvalues, orthogonal linear combinations can be formed.

Definition. The largest possible number of mutually orthogonal linear combinations of eigenfunctions of an operator Q having the same eigenvalue q is said to be the **degree of degeneracy of** q.

There is a systematic procedure to form such orthogonal linear combinations, the so-called **Schmidt orthogonalization procedure.** In the orthogonalization of ψ_i and ψ_k, above, we have used the Schmidt procedure for the special case of twofold degeneracy.

6.2   MORE ABOUT EXPECTATION VALUES

We shall now take another, closer, look at expectation values. Let Q be a hermitian operator with a complete set of normalized eigenfunctions u_k, $k = 1, 2, 3, \ldots$ belonging to the eigenvalues q_k, $k = 1, 2, 3, \ldots$. Let a quantum mechanical system be in a state described by a wave function ψ_n which *is not* an eigenfunction to Q. The expectation value of Q is then

$$\langle Q \rangle_n = (\psi_n, Q\psi_n) \tag{6.44}$$

We expand ψ_n:

$$\psi_n = \sum_{i=1}^{\infty} a_{ni} u_i \tag{6.45}$$

Substitution into Eq. 6.44 yields

$$\langle Q \rangle_n = \left(\sum_{i=1}^{\infty} a_{ni} u_i, Q \sum_{k=1}^{\infty} a_{nk} u_k \right) = \left(\sum_{i=1}^{\infty} a_{ni} u_i, \sum_{k=1}^{\infty} a_{nk} q_k u_k \right) \tag{6.46a}$$

since $(u_i, u_k) = \delta_{ik}$, this becomes

$$\langle Q \rangle_n = \sum_{i=1}^{\infty} a_{ni}{}^* a_{ni} q_i = \sum_{i=1}^{\infty} |a_{ni}|^2 q_i \tag{6.46b}$$

In other words, the expectation value of Q in the state ψ_n is the weighted average of all the eigenvalues q_i, and the statistical weight of each eigenvalue is the absolute square of the expansion coefficient of the corresponding eigenfunction of Q.

We see now what the physical significance of the eigenfunctions of an operator is: If the wave function of the system is an eigenfunction of Q with the eigenvalue q_m, the statistical weights of all the other eigenvalues are zero and a measurement will yield q_m with certainty. But can we square this statement with the uncertainty principle? We certainly can. The uncertainty principle links the uncertainties of two observables and says nothing about the uncertainty in the measurement of a single observable. However, not all pairs of observables are linked with an uncertainty relation of the type of equation Eq. 2.36. In view of the above statement about eigenfunctions, we can pronounce the following theorem:

THEOREM

If the wave function of a quantum mechanical system is an eigenfunction of the operator A and at the same time an eigenfunction of the operator B, we can simultaneously measure $\langle A \rangle$ and $\langle B \rangle$ with certainty.

Proof. The proof follows from the foregoing statement about eigenfunctions. To find out under what circumstances a wave function can be an eigenfunction of two or more operators, we make another brief excursion into mathematics.

Definition. ψ is said to be a **simultaneous eigenfunction** of the linear operators A and B, belonging to the eigenvalues α and β respectively if it satisfies both:

$$A\psi = \alpha\psi \quad \text{and} \quad B\psi = \beta\psi \tag{6.47}$$

THEOREM

Two linear operators A and B which have simultaneous eigenfunctions, ψ, commute.

Proof

$$A(B\psi) = A\beta\psi = \beta A\psi = \beta\alpha\psi \qquad (6.48)$$

$$B(A\psi) = B\alpha\psi = \alpha B\psi = \alpha\beta\psi \qquad (6.49)$$

hence

$$BA\psi = AB\psi \qquad \text{q.e.d.} \qquad (6.50)$$

We show now that the inverse statement also holds true.

THEOREM

The eigenfunctions of commuting linear operators are simultaneous eigenfunctions or in the case of degeneracy can be constructed in such a way as to be simultaneous eigenfunctions.

Proof

(a) *Nondegenerate case.* Let ψ be an eigenfunction to A

$$A\psi = \alpha\psi \qquad (6.51)$$

We multiply both sides *from the left* with B

$$BA\psi = B\alpha\psi \qquad (6.52)$$

since A and B are assumed to commute

$$BA\psi = AB\psi = \alpha B\psi \qquad (6.53)$$

i.e., $B\psi$ is an eigenfunction of A with the eigenvalue α. Since α was assumed to be nondegenerate, $B\psi$ can only be a multiple of ψ, hence

$$B\psi = \beta\psi \qquad \text{q.e.d.} \qquad (6.54)$$

(b) *Degenerate case.* For the sake of simplicity we assume α to be twofold degenerate and ψ_1 and ψ_2 to be two orthonormal eigenfunctions of A belonging to α. Then all linear combinations of ψ_1 and ψ_2 are also eigenfunctions of A with the eigenvalue α. We pick one of them:

$$\psi = a_1\psi_1 + a_2\psi_2 \qquad (6.55)$$

It must be an eigenfunction of A:

$$A\psi = \alpha(a_1\psi_1 + a_2\psi_2) \qquad (6.56)$$

We multiply Eq. 6.56 *from the left* with B and obtain

$$BA\psi = \alpha B(a_1\psi_1 + a_2\psi_2) \qquad (6.57)$$

Since A and B commute, this can also be written

$$AB\psi = \alpha B(a_1\psi_1 + a_2\psi_2) \qquad (6.58)$$

In other words, $B(a_1\psi_1 + a_2\psi_2)$ is an eigenfunction of A with the eigenvalue α. Since α was assumed to be twofold degenerate, $B(a_1\psi_1 + a_2\psi_2)$ must be one of the linear combinations of ψ_1 and ψ_2. We must thus be able to write

$$B(a_1\psi_1 + a_2\psi_2) = c_1\psi_1 + c_2\psi_2 \qquad (6.59)$$

Since the coefficients a_1 and a_2 are still arbitrary we can determine them so that

$$c_1\psi_1 + c_2\psi_2 = \beta(a_1\psi_1 + a_2\psi_2) \qquad (6.60)$$

In this case $a_1\psi_1 + a_2\psi_2$ also becomes an eigenfunction of B since it satisfies

$$B(a_1\psi_1 + a_2\psi_2) = \beta(a_1\psi_1 + a_2\psi_2) \qquad (6.61)$$

To determine the particular coefficients a_1 and a_2 that satisfy Eq. 6.60, we multiply Eq. 6.61 from the left with ψ_1^* and ψ_2^* respectively and integrate. This yields

$$a_1(\psi_1, B\psi_1) + a_2(\psi_1, B\psi_2) = \beta a_1(\psi_1, \psi_1) + \beta a_2(\psi_1, \psi_2) \qquad (6.62)$$

and

$$a_1(\psi_2, B\psi_1) + a_2(\psi_2, B\psi_2) = \beta a_1(\psi_2, \psi_1) + \beta a_2(\psi_2, \psi_2) \qquad (6.63)$$

ψ_1 and ψ_2 were assumed to be orthonormal, hence, if we use the abbreviations

$$(\psi_1, B\psi_1) = B_{11}, \qquad (\psi_1, B\psi_2) = B_{12}, \qquad , \text{ etc.}$$

$$a_1 B_{11} + a_2 B_{12} = \beta a_1 \qquad (6.64)$$

and

$$a_1 B_{21} + a_2 B_{22} = \beta a_2 \qquad (6.65)$$

These are two simultaneous linear equations for a_1 and a_2. They can be solved if the determinant of their coefficients vanishes.

$$\begin{vmatrix} B_{11} - \beta & B_{12} \\ B_{21} & B_{22} - \beta \end{vmatrix} = 0 \qquad (6.66)$$

This is a quadratic equation with two roots β_1 and β_2. We had assumed to know B, ψ_1, and ψ_2 and, hence, we can calculate B_{11}, B_{12}, B_{21}, B_{22} and, thereby, β_1 and β_2. If β_1 and β_2 are equal, any choice of a_1 and a_2 will satisfy Eqs. 6.64 and 6.65, and β is a degenerate eigenvalue. If $\beta_1 \neq \beta_2$ Eqs. 6.64 and 6.65 will determine the constants a_1 and a_2 so that Eq. 6.61 is satisfied, q.e.d.

Obviously this procedure can be extended to higher degrees of degeneracy. Returning from this mathematical digression to the realm of physics, we can formulate the following theorem.

THEOREM

The expectation values of two commuting operators can be measured simultaneously and with arbitrary precision.[8]

6.3 COMMUTATORS AND THE UNCERTAINTY PRINCIPLE

In view of the preceding theorem the uncertainty principle must be restricted to the expectation values of noncommuting operators. To investigate this situation in detail we avail ourselves of some mathematical tools. If two operators A and B do not commute, the difference of the products AB and BA must be different from zero.

Since expressions of the type $(AB - BA)$ occur frequently in quantum mechanics, this difference has been given a special name and a special symbol.

Definition

$$AB - BA = [A, B] \tag{6.67}$$

is called the **commutator** of the operators A and B.

From this definition follow, immediately, the identities below which the reader may verify himself.

$$[A, B] \equiv -[B, A] \tag{6.68}$$

$$[A, BC] \equiv [A, B]C + B[A, C] \tag{6.69}$$

$$[AB, C] \equiv [A, C]B + A[B, C] \tag{6.70}$$

$$[A, [B, C]] + [B, [C, A]] + [C, [A, B]] \equiv 0 \tag{6.71}$$

The commutator of two operators A and B will in the most general case be another operator, for example, D.

$$[A, B] = D \tag{6.72}$$

We assume now that A and B are hermitian and investigate the hermiticity—or the lack of it—of D.

$$(\psi, [A, B]\varphi) = (\psi, AB\varphi) - (\psi, BA\varphi) = (\psi, D\varphi) \tag{6.73}$$

We compare this with

$$(D\psi, \varphi) = (AB\psi, \varphi) - (BA\psi, \varphi) \tag{6.74}$$

---

[8] At least in principle, that is.

Since A and B are hermitian this can be written

$$(D\psi, \varphi) = (B\psi, A\varphi) - (A\psi, B\varphi) = (\psi, BA\varphi) - (\psi, AB\varphi) = -(\psi, D\varphi)$$
(6.75)

An operator satisfying Eq. 6.75 is said to be **antihermitian,** hence:

THEOREM

The commutator of two hermitian operators is antihermitian. If we want to express Eq. 6.72 in terms of a hermitian operator we can write

$$[A, B] = iC$$
(6.76)

where C is hermitian. We introduce now a new operator

$$Q = A + i\lambda B$$
(6.77)

and form

$$(Q\psi, Q\psi) = \int |Q\psi|^2 \, d\tau \geqslant 0$$
(6.78)

substitution of Eq. 6.77 into Eq. 6.78 yields

$$((A + i\lambda B)\psi, (A + i\lambda B)\psi) \geqslant 0$$
(6.79)

Since A and B are hermitian this can be written as[9]

$$(\psi, (A - i\lambda B)(A + i\lambda B)\psi) \geqslant 0$$
(6.80)

Making use of Eq. 6.76, this can also be written as

$$\langle A^2 \rangle - \lambda \langle C \rangle + \lambda^2 \langle B^2 \rangle \geqslant 0$$
(6.81)

This inequality must hold regardless of the size of λ, and we ask for which value of λ the left side becomes smallest. Obviously for

$$\lambda = \frac{\langle C \rangle}{2\langle B^2 \rangle}$$
(6.82)

assuming that $\langle B^2 \rangle \neq 0$. Using Eq. 6.82 in Eq. 6.81, we get

$$\langle A^2 \rangle - \frac{\langle C \rangle^2}{2\langle B^2 \rangle} + \frac{\langle C \rangle^2}{4\langle B^2 \rangle} \geqslant 0$$
(6.83)

or

$$\langle A^2 \rangle \langle B^2 \rangle \geqslant \frac{\langle C \rangle^2}{4}$$
(6.84)

The heading of this paragraph may have led the reader to suspect that this expression might be related to the uncertainty principle, and it is. To show

[9] Note that since B is hermitian, iB is antihermitian and changes its sign in our notation if it is moved to operate on the second function in the bracket.

this we introduce an operator that describes the deviation of an individual measurement of a quantity described by the operator A from its expectation value:

$$\delta A = A - \langle A \rangle \tag{6.85}$$

The expectation value of this operator is zero, as it must be, since the expectation value $\langle A \rangle$ is *defined* as the average of many measurements of the quantity represented by the operator A.

In an investigation of experimental errors it is customary to characterize the accuracy of a series of measurements by the "root-mean-square" (rms) error[10] of the measurements. In our case the root-mean-square error is described by the operator

$$(\delta A)^2 = A^2 - 2A\langle A \rangle + \langle A \rangle^2 \tag{6.86}$$

Its expectation value—which can be different from zero even if $\langle \delta A \rangle = 0$—is

$$\langle (\delta A)^2 \rangle = \langle A^2 \rangle - 2\langle A \rangle^2 + \langle A \rangle^2 \tag{6.87}$$

or

$$\langle (\delta A)^2 \rangle = \langle A^2 \rangle - \langle A \rangle^2 \tag{6.88}$$

Now we replace A with $\delta A = A - \langle A \rangle$ and B with $\delta B = B - \langle B \rangle$ on both sides of the inequality Eq. 6.84. This gives on the left-hand side:

$$\langle (\delta A)^2 \rangle \langle (\delta B)^2 \rangle$$

The right-hand side of Eq. 6.84 remains unchanged since Eq. 6.85 and its equivalent for B obviously have the same commutator as A and B themselves. Hence we obtain

$$\langle (\delta A)^2 \rangle \langle (\delta B)^2 \rangle \geqslant \frac{\langle C \rangle^2}{4} \tag{6.89}$$

or, since

$$\Delta A = \sqrt{\langle (\delta A)^2 \rangle} \quad \text{and} \quad \Delta B = \sqrt{\langle (\delta B)^2 \rangle} \tag{6.90}$$

are the rms errors of the measurements of the properties described by the operators A and B, we find

$$\Delta A \, \Delta B \geqslant \frac{\langle C \rangle}{2} \tag{6.91}$$

---

[10] Let x be a quantity to be measured, x_k the result of an individual measurement, and $\bar{x} = \frac{1}{n} \sum_1^n x_k$ the average, or mean, of all measurements. $\delta x_k = \bar{x} - x_k$ is the deviation of the kth measurement from the mean.

$$\langle \delta x_k \rangle = \sqrt{\frac{1}{n} \sum_1^n (\bar{x} - x_k)^2}$$

is the rms error. In the case of a gaussian distribution of the x_k around $\bar{x}$, the rms error gives the turning point of the Gaussian.

i.e., *if the operators describing two properties of a quantum-mechanical system do not commute, the product of the uncertainties in the measurement of both quantities is larger than or equal to a certain minimum value.* This is the most general form of the **uncertainty principle.** In most cases of interest the commutator of two noncommuting hermitian operators will not be another operator but a constant.

As an example we calculate the uncertainty product of position and momentum

$$[x, p_x] = \left[x, -i\hbar \frac{\partial}{\partial x} \right] = -i\hbar \left[x, \frac{\partial}{\partial x} \right] \tag{6.92}$$

according to Appendix A.1.2

$$\left[x, \frac{\partial}{\partial x} \right] = -1 \tag{6.93}$$

Hence, in this case, $C = \hbar$ or

$$\Delta x \, \Delta p_x \geqslant \frac{\hbar}{2} \tag{6.94}$$

We notice that the uncertainty product given by Eq. 6.94 is smaller by a factor of $1/4\pi$ than the value given by Eq. 1.9. This is due to the fact that in the present calculation we had *minimized* the uncertainty product whereas the previous values were only rough estimates. The value given by Eq. 6.94 is, accordingly, said to be the **minimum uncertainty product.**[11]

6.4 ANGULAR MOMENTUM COMMUTATORS

Since H, L^2, and L_z have simultaneous eigenfunctions (namely, the hydrogen eigenfunctions) they must commute, i.e.,

$$[H, L^2] = [H, L_z] = [L^2, L_z] = 0 \tag{6.95}$$

To investigate the remaining components of the angular momentum operator we form[12]

$$[L_x, L_y] = (yp_z - zp_y)(zp_x - xp_z) - (zp_x - xp_z)(yp_z - zp_y) \tag{6.96}$$

Since

$$[p_x, p_y] = [x, p_y] = 0, \text{ etc.,}$$

this becomes

$$[L_x, L_y] = yp_xp_zz - yxp_z{}^2 - z^2p_yp_x + xp_yzp_y - yp_xzp_z + z^2p_xp_y$$
$$+ yxp_z{}^2 - xp_yp_zz = [z, p_z](xp_y - yp_z) \tag{6.97}$$

---

[11] See also Eq. 2.36.

[12] p_x, p_y, and p_z are used as abbreviations for the components of the *momentum operator* $-i\hbar(\partial/\partial x)$, $-i\hbar(\partial/\partial y)$, and $-i\hbar(\partial/\partial z)$.

Now

$$[z, p_z] = \left[z, -i\hbar \frac{\partial}{\partial z}\right] = -i\hbar\left[z, \frac{\partial}{\partial z}\right] \qquad (6.98)$$

According to Example A.1.1 in Appendix A.1, we have

$$\left[z, \frac{\partial}{\partial z}\right] = -1 \qquad \text{or} \qquad [z, p_z] = i\hbar \qquad (6.99)$$

and hence, using Eq. 5.89, Eq. 6.97 becomes

$$[L_x, L_y] = i\hbar L_z \qquad (6.100)$$

Through a permutation of the variables we also obtain

$$[L_y, L_z] = i\hbar L_x \qquad (6.101)$$

and

$$[L_z, L_x] = i\hbar L_y \qquad (6.102)$$

Equation 6.95 might create the impression that the L_z enjoys some sort of preferred position; this is, of course, not the case. From

$$[L^2, L_z] = [L_x^2 + L_y^2 + L_z^2, L_z] = 0 \qquad (6.103)$$

it follows immediately that

$$[L^2, L_x] = 0 \qquad (6.104)$$

and

$$[L^2, L_y] = 0 \qquad (6.105)$$

through a permutation of the variables. The fact that Eq. 5.106 looks so much simpler than the corresponding expressions for L_x and L_y is a result of the particular choice of the spherical polar coordinate system. Since the Hamiltonian is symmetrical in the variables x, y, and z we can also conclude that

$$[H, L_x] = [H, L_y] = 0 \qquad (6.106)$$

As Eq. 6.97 demonstrates, it does not follow from Eq. 6.106 that L_x and L_y commute. From the fact that H commutes with L_x, L_y, and L_z whereas L_x, L_y, and L_z do not commute with each other, we have to conclude that the simultaneous eigenfunctions which H has with L_x differ from those it has with L_y and L_z.

We wish to draw attention to the fact that all the above conclusions were drawn directly from the algebraic properties of the angular momentum commutators. No reference had to be made to any specific set of eigenfunctions.

6.5 THE TIME DERIVATIVE OF AN EXPECTATION VALUE

We shall now derive a useful expression for the time derivative of an expectation value and prove an important theorem.

Let Q be an operator describing some property of a quantum-mechanical system. We assume Q to have no explicit time dependence, i.e.,

$$\frac{\partial Q}{\partial t} = 0 \tag{6.107}$$

and ask: What is the rate of change of the expectation value of Q?

$$\frac{d}{dt} \langle Q \rangle = \int \frac{\partial \psi^*}{\partial t} Q \psi \, d\tau + \int \psi^* Q \frac{\partial \psi}{\partial t} \, d\tau \tag{6.108}$$

For $\partial \psi / \partial t$ and $\partial \psi^* / \partial t$ we substitute the appropriate expressions from the time-dependent Schrödinger equation and its complex conjugate:

$$i\hbar \frac{\partial \psi}{\partial t} = H\psi \quad \text{and} \quad -i\hbar \frac{\partial \psi^*}{\partial t} = H\psi^* \tag{6.109}$$

(H itself is real.) Hence

$$\frac{d}{dt} \langle Q \rangle = \frac{1}{i\hbar} \int (-H\psi^* Q\psi + \psi^* QH\psi) \, d\tau \tag{6.110}$$

Or, using our new notation,

$$\frac{d}{dt} \langle Q \rangle = \frac{1}{i\hbar} \{ -(H\psi, Q\psi) + (\psi, QH\psi) \} \tag{6.111}$$

We have shown in Chapter 6.1 that H is hermitian. Hence[13]

$$(H\psi, Q\psi) = (\psi, HQ\psi) \tag{6.112}$$

and therefore

$$\frac{d}{dt} \langle Q \rangle = \frac{1}{i\hbar} \{ (\psi, QH\psi) - (\psi, HQ\psi) \} = \frac{1}{i\hbar} (\psi, [Q, H]\psi) \tag{6.113}$$

If $[Q, H] = 0$ it follows obviously that $(d/dt)\langle Q \rangle = 0$. We can, thus, formulate the following important **theorem:**

The expectation value of an operator that commutes with the Hamiltonian is constant in time.

Obviously this is a sufficient but not a necessary condition since Eq. 6.113 can also vanish if $[Q, H]\psi$ is orthogonal to ψ.

---

[13] To see that this is nothing but the definition of hermiticity, just rename the function $Q\psi$ by letting $Q\psi = \varphi$.

6.6 SEPARABILITY AND CONSERVATION LAWS

In Chapter 5.9 we made a then cryptic remark about the connection between the separability of the Schrödinger equation and the conservation of angular momentum. We shall now try to shed some more light on this connection. Let H be an operator, for example, the Hamiltonian of the hydrogen atom. Let the eigenvalue equation

$$H\psi = E\psi \tag{6.114}$$

i.e., the Schrödinger equation, be separable. This means that Eq. 6.114 is equivalent to

$$H_1\psi_1 = E_1\psi_1; \qquad H_2\psi_2 = E_2\psi_2 \tag{6.115}$$

where

$$H_1 + H_2 = H, \ \psi_1\psi_2 = \psi \qquad \text{and} \qquad E_1 + E_2 = E$$

H_1 and ψ_1 contain only variables not contained in H_2 and ψ_2. Multiplying Eq. 6.115 with ψ_2 and ψ_1, respectively, we obtain

$$\psi_2 H_1\psi_1 = H_1\psi = E_1\psi \qquad \text{and} \qquad \psi_1 H_2\psi_2 = H_2\psi = E_2\psi \tag{6.116}$$

This means that the eigenfunction ψ is a simultaneous eigenfunction of the Hamiltonian H *and* its two parts H_1 and H_2. According to the theorem on p. 113 this implies that H_1, H_2 and H commute:

$$[H_1, H_2] = [H, H_1] = [H, H_2] = 0 \tag{6.117}$$

This, in turn, means, according to Chapter 6.5, that the properties described by the operators H_1 and H_2 are constant in time. We have, thus, shown that the separability of the Hamiltonian implies the conservation of some dynamic variable. The well-known conservation of angular momentum is, thus, quantum mechanically expressed by the fact that the Hamiltonian *can* be written as the sum of a radial part and the angular momentum operator. We have said "*can* be written," since separability is a *sufficient* but *not a necessary* condition for the conservation of dynamic variables. The separability of a partial differential equation is dependent on the coordinate system whereas physical conservation laws are not. As an example we consider the angular momentum operator L^2. In spherical polar coordinates we saw this operator appear explicitly in the Hamiltonian, enabling us to separate the Schrödinger equation into an angular and a radial part. In cartesian coordinates this would not have been the case. The angular momentum operator would have been dependent on all three coordinates, and its explicit introduction into the Schrödinger equation would not have led to a separation of the variables. The choice of spherical polar coordinates with all its fortunate consequences is, of course, not a stroke of luck but a tribute to the insight of Erwin Schrödinger.

6.7 PARITY

We had seen earlier that the eigenfunctions of the harmonic oscillator have a definite parity. The same is true for the hydrogen eigenfunctions. Therefore, it might be appropriate to track down the origin of this peculiar property of certain eigenfunctions. To this end we consider the Schrödinger equation:

$$Eu(\mathbf{r}) = -\frac{\hbar^2}{2m} \nabla^2 u(\mathbf{r}) + V(\mathbf{r})u(\mathbf{r}) \qquad (6.118)$$

Changing the sign of all coordinates yields

$$Eu(-\mathbf{r}) = -\frac{\hbar^2}{2m} \nabla^2 u(-\mathbf{r}) + V(-\mathbf{r})u(-\mathbf{r}) \qquad (6.119)$$

if

$$V(-\mathbf{r}) = V(\mathbf{r})$$

i.e., if the potential is invariant under space inversion, the two differential equations are identical. This means that $u(\mathbf{r})$ and $u(-\mathbf{r})$ can differ only by a constant factor α that is not determined by the differential equation. Hence

$$u(-\mathbf{r}) = \alpha u(\mathbf{r}) \qquad (6.120)$$

Changing the sign of $\mathbf{r}$ we get

$$u(\mathbf{r}) = \alpha u(-\mathbf{r}) = \alpha^2 u(\mathbf{r}) \qquad (6.121)$$

Hence

$$\alpha^2 = 1 \qquad \text{or} \qquad \alpha = \pm 1$$

In other words, *if the potential is symmetrical* about the origin, the eigenfunctions have a definite parity. They will either change sign (odd parity) or not change sign (even parity) under a mirror transformation.

We still have to plug one loophole. Our proof depended on the assumption that the eigenfunction $u(\mathbf{r})$ is nondegenerate. This is not always the case. Therefore, we assume now that the Schrödinger equation Eq. (6.114) has several degenerate eigenfunctions and that one of these eigenfunctions, properly normalized, is $u(\mathbf{r})$; $u(\mathbf{r})$ need, of course, not have a definite parity.

Through a process called symmetrization, or antisymmetrization, we can construct from $u(\mathbf{r})$ normalized eigenfunctions of even, or odd, parity.

The equation

$$u_e(\mathbf{r}) = \gamma[u(\mathbf{r}) + u(-\mathbf{r})] \qquad (6.122)$$

is obviously an eigenfunction, it has even parity, and it can be normalized through the choice of γ. Similarly

$$u_0(\mathbf{r}) = \gamma'[u(\mathbf{r}) - u(-\mathbf{r})] \qquad (6.123)$$

has odd parity and can be normalized through the choice of γ'. We sum up: If the potential in the Hamiltonian is symmetrical the eigenfunctions of a nondegenerate eigenvalue have a definite even, or odd, parity. In the case of degeneracy we can always construct eigenfunctions of definite parity through symmetrization or antisymmetrization.

We now show that the parity of the wave function of a system is constant in time. To this end we invent an operator P that consists of the instruction: *change the sign of all coordinates in the function that follows.* This operator applied to an eigenfunction of even parity

$$Pu_e(\mathbf{r}) = u_e(-\mathbf{r}) = u_e(\mathbf{r}) \qquad (6.124)$$

has the eigenvalue $+1$. Applied to an eigenfunction of odd parity

$$Pu_0(\mathbf{r}) = u_0(-\mathbf{r}) = -u_0(\mathbf{r}) \qquad (6.125)$$

it has the eigenvalue -1. In other words, the eigenvalue of this operator is the parity of the wave function. Hence

$$\langle P \rangle = \begin{cases} +1 & \text{for even parity eigenfunctions} \\ -1 & \text{for odd parity eigenfunctions} \end{cases}$$

This operator is, therefore, called the **parity operator.** We apply the parity operator to the Schrödinger equation. This yields

$$P[Hu(\mathbf{r})] = Hu(-\mathbf{r}) = HPu(\mathbf{r}) = Eu(-\mathbf{r}) \qquad (6.126)$$

if the Hamiltonian

$$H = -\frac{\hbar^2}{2m}\nabla^2 + V(\mathbf{r}) \qquad (6.127)$$

is symmetrical in $\mathbf{r}$, which it is as long as $V(\mathbf{r})$ is symmetrical. Another way to express the statement made by Eq. 6.126 is to say that

$$PH = HP \qquad \text{or} \qquad [H, P] = 0 \qquad (6.128)$$

According to Eq. 6.113 this implies

$$\frac{d}{dt}\langle P \rangle = 0 \qquad (6.129)$$

This digression about the parity of eigenfunctions might seem rather academic, and it is *at this point.* The true importance of the concept of parity will become evident later in two applications:

1. In complicated quantum-mechanical systems it would sometimes require laborious calculations to tell whether a certain process can happen. If we know, however, that the **interaction Hamiltonian**[14] of the system is symmetrical and that the process would change the parity of the wave function, we know immediately that it cannot happen.

2. If we do not know the interaction Hamiltonian[14] (and we do not in the case of nuclear interactions), we can measure whether certain processes conserve parity and, thus, learn whether the Hamiltonian is symmetrical. This may be less than we would like to learn but it is often what we must settle for.

PROBLEMS

6.1 Do the functions log (kx), $k = 0, 1, 2, \ldots$ form a complete set? Do the functions log (kx), $k = \cdots -2, -1, 0, 1, 2, \ldots$ form a complete set?

6.2 Show that if A is a hermitian operator, A^2 is also hermitian.

6.3 A and B are two arbitrary hermitian operators. Which of the following expressions are hermitian: $[A, B]$, $i[A, B]$, AB?

6.4 Show that the operator of the z-component of the angular momentum is hermitian.

6.5 Can we measure the energy and the momentum of a particle simultaneously with arbitrary precision?

6.6 Can we measure the kinetic and the potential energy of a particle simultaneously with arbitrary precision?

6.7 A quantum-mechanical particle moves in a constant potential (constant in time and space). Is its potential energy constant in time?

6.8 Express the degree of degeneracy of the hydrogen energy eigenvalues as a function of n.

6.9 Verify Eqs. 6.69, 6.70, and 6.71.

6.10 Are the expectation values of the kinetic energy and the potential energy of a bound particle constant in time?

6.11 Find the parity of the $1s$, $2s$, $2p$, $3s$, $3p$, and $3d$ eigenfunctions of the hydrogen atom.

SOLUTIONS

6.5 In order for E and p to be simultaneously measurable, their operators must commute. Now

$$H = -\frac{\hbar^2}{2m} \nabla^2 + V \quad \text{and} \quad \mathbf{p} = -i\hbar \nabla$$

---

[14] The word Hamiltonian is not only used for the operator of the total energy of a stationary state but also for the operator of the interaction energy when a system goes from one state to another. It is this latter operator that we refer to with the term interaction Hamiltonian.

Hence, since $\nabla^2 V = \nabla V^2$,

$$[H, \mathbf{p}] = -i\hbar(V\nabla - \nabla V)$$

This commutator vanishes only in the case $V = $ constant. In other words, energy and momentum can be measured simultaneously and with arbitrary precision only for an unbound particle. This is, of course, exactly what we would expect from the uncertainty principle. Given enough time we can measure the energy of a bound particle with arbitrary precision. The accuracy of the measurement of the momentum is, on the other hand, limited by the fact that a bound particle is localized.

6.11 We investigate here the parity of the eigenfunction $u(n, l, m) = u(3, 2, 1)$ leaving it to the reader to find a general connection between parity and the quantum numbers n, l, and m. First we must find out how the variables r, ϑ, and φ change as we change $x \to -x$, $y \to -y$, and $z \to -z$. Obviously r remains unchanged. Consulting Figure 5.2, we find that a vector pointing in a certain direction will point in the opposite direction if we replace ϑ with $\pi - \vartheta$ *and* φ with $\pi + \varphi$. Making these substitutions in the $3d$ wave function, we find that

$$\sin \vartheta \cos \vartheta \, e^{i\varphi} \to \sin (\pi - \vartheta) \cos (\pi - \vartheta) e^{i\varphi} \cdot e^{i\pi}$$
$$= \sin \vartheta \, (-\cos \vartheta) e^{i\varphi} \cdot (-1) = \sin \vartheta \cos \vartheta \, e^{i\varphi}$$

i.e., the parity of the $u(3, 2, 1)$ wave function is even.

7

ATOMIC PHYSICS

In classical mechanics we can solve the two-body problem in closed form. For more complicated problems, however, we must resort to perturbation theory. A similar situation exists in quantum mechanics. We have seen the hydrogen problem yield to a frontal assault. Many of the more complicated problems, however, require the use of approximative methods. Several such methods have been developed, using either iterative procedures or a statistical approach. The more important iterative methods, first introduced by **D. R. Hartree,** assume that each electron moves in a *central* field created by the nucleus and all the other electrons. The Schrödinger equation is solved for each electron in an assumed central field, and the thus-found electron eigenfunctions give a charge distribution. This charge distribution is then used to calculate new eigenfunctions which in turn give a charge distribution that is somewhat closer to reality. This new charge distribution is then used to calculate new eigenfunctions which in turn . . . .

Today the Hartree method, applied by high-speed computers, allows us to calculate numerically the electron distribution in many electron atoms with high precision. While Hartree calculations are not overly difficult conceptually, their application to real physical problems requires a truly heroic effort.

Even if we do not endeavor to calculate the eigenfunctions and energy levels of the more complicated atoms, their fine structure offers a fascinating field of study. The fine structure of spectral lines results from the existence of degenerate eigenvalues whose various eigenfunctions react differently to perturbations. We shall discuss this in some detail in a later chapter dealing with perturbation theory. The degeneracies are, of course, always connected with different angular momentum states, and a quantitative treatment requires a thorough knowledge of the quantum-mechanical theory of angular momentum. This theory is based on the algebraic properties of the angular momentum operators, some of which we derived in Chapter 6.4. A detailed theory of angular momentum is beyond the scope of this book, although we shall derive some of its conclusions in Chapters 9 and 10.

126

Thus we conclude that a systematic discussion of the spectra of many electron atoms is out of our reach. In this chapter, therefore, we shall try, in a less strenuous way, to come to a qualitative understanding of some of the features of many electron atoms by combining experimental results, basic quantum mechanics, and educated guesses. Our motivation for this chapter is threefold: (1) to furnish a nodding acquaintance with the important field of atomic physics, (2) to introduce the concept of spin, and (3) to introduce the terminology and notation of atomic physics.

7.1 THE PERIODIC TABLE

Even before the advent of quantum mechanics the striking similarities and regularities in the chemical behavior of the elements were attributed to regularities in their atomic structure. From our quantum-mechanical vantage point we expect these regularities to reflect the structure of the wave functions of the atomic electrons. The electron structure of all the known elements has now been unraveled, using mainly spectroscopic evidence. In the following we shall trace the assignment of quantum numbers for the lightest elements.[1] The ionization energy of the hydrogen atom is

$$E_{\text{ion}} = \frac{me^4}{2\hbar^2} \tag{5.80}$$

According to Eq. 5.84 this becomes for a *one electron atom* of nuclear charge Ze

$$E_{\text{ion}} = \frac{me^4 Z^2}{2\hbar^2} \tag{7.1}$$

We apply this to the He atom.

Helium

In the He atom the nucleus has a charge $2e$; however, each of the two electrons neutralizes part of this charge for the other electron so that none of them sees the full nuclear charge. Let us assume that one of the electrons is in a $1s$ state. This electron together with the nucleus presents a spherical charge distribution to the second electron, and we assume that the second electron moves at least approximately as it would in the field of a central charge eZ_{eff} somewhere between e and $2e$. Assuming that the second electron is also in a $1s$ state we calculate the ionization energy of the He atom.

$$E_{\text{ion}} = \frac{me^4 Z_{eff}^2}{2\hbar^2} \tag{7.2}$$

---

[1] Following a treatment given in W. Weizel, *Lehrbuch der Theoretischen Physik* 2nd volume, Springer Verlag, Berlin, 1950.

Equating this to the measured value $E_{\text{ion}} = 24.5$ eV (the ionization energy of hydrogen is 13.5 eV) we find

$$Z_{eff} = \sqrt{\frac{24.5}{13.5}} = 1.35 \qquad (7.3)$$

This means that each of the electrons sees only $1.35e$ instead of $2e$, or that one electron screens 65 percent of one proton charge. This is not unreasonable.

Lithium

Assuming that the principal quantum number of the third electron is also $n = 1$, we obtain, using the measured value $E_{\text{ion}} = 5.37$ eV,

$$Z_{eff} = \sqrt{\frac{5.36}{13.5}} = 0.63 \qquad (7.4)$$

This is impossible since two electrons cannot neutralize 2.37 proton charges. We try $n = 2$ for the third electron. This yields

$$Z_{eff} = \sqrt{\frac{5.36}{3.38}} = 1.26 \qquad (7.5)$$

(3.38 eV is the ionization energy of hydrogen in its $2s$ state.) This means that, seen from a $2s$ electron, each of the two $1s$ electrons screens $1.74/2 = 87$ percent of a proton charge. This is reasonable since the $2s$ electron is, on the average, farther away from the nucleus than the two $1s$ electrons.

Beryllium

We assume $n = 2$ for the fourth electron and also assume that each of the two $1s$ electrons screens 87 percent of one proton charge, just as they did for the first $2s$ electron. In addition, the second $2s$ electron is also screened from the nucleus by the first $2s$ electron.

In the case of He, we observed that two electrons with equal n screen each other with 65 percent efficiency, and we now assume that the same holds true for the two $2s$ electrons. This leaves us with

$$Z_{eff} = 4 - 1.74 - 0.65 = 1.61 \qquad (7.6)$$

For $n = 2$ this yields an ionization energy of

$$E_{\text{ion}} = 3.38 \cdot (1.61)^2 = 8.8 \text{ eV} \qquad (7.7)$$

and this is in fair agreement with the experimental value of 9.28 eV. We can, thus, assume that the assigned quantum numbers are correct.

Boron

Assuming another $2s$ electron, we calculate from the known ionization energy $E_{\text{ion}} = 8.26$ eV that $Z_{eff} = 1.56$. If we calculate Z_{eff}, assuming that the screening due to the 4*th* electron is 65 percent, we get

$$Z_{eff} = 5 - 1.74 - 1.3 = 1.96 \tag{7.8}$$

which is a rather poor agreement. So we try $n = 3$. This yields $Z_{eff} = 2.34$ which is impossible, since it would mean that an $n = 3$ electron (which is, on the average, farther away from the nucleus than an $n = 2$ electron) sees more of the nuclear charge.

This suggests that the 5th electron might be a $2p$ electron. The $2p$ eigenfunction is zero at the origin, and the screening should, therefore, be better than the 3.04 nuclear charges Eq. 7.8 that we expected for another $2s$ electron.

If we continue this procedure, adding evidence from optical and x-ray spectra at the later stages, we arrive at the quantum number assignments listed in Table 7.1.

7.2   THE PAULI PRINCIPLE

Table 7.1 reveals some remarkable regularities. There are never more than two *s*-electrons, six *p*-electrons, ten *d*-electrons, or fourteen *f*-electrons belonging to the same value of the principal quantum number n. Remembering that a *p*-state is threefold degenerate, a *d*-state fivefold degenerate, etc., we conjecture that the six *p*-electrons represent all three possible values of m $(-1, 0, +1)$ just twice, etc. Thus we formulate, *tentatively* the following theorem:

There are never more than two electrons in any atom that have the same set of quantum numbers.

There are never more than two electrons in any atom that have the same set of quantum numbers.

This looks suspicious. Why should there be just two electrons allowed to a state? It was in 1925 when **Wolfgang Pauli** suggested that there might be another quantum number at the bottom of this.[2] This quantum number was to be capable of only two values, and two electrons of equal quantum numbers n, l, and m were to differ in this new quantum number. Soon afterward this quantum number was identified by Uhlenbeck and Goudsmit as the "spin" of the electron.[3] Later, we shall discuss the spin in more detail. Here all we need to know is that it exists and that, in the case of the electron, it gives rise to a quantum number capable of only two values. This enables us to pronounce the famous Pauli exclusion principle or **Pauli principle**:

"There are no two electrons in any atom that have the same set of quantum numbers."

[2] W. Pauli, *ZS. f. Phys.*, **31**, 373 (1925).
[3] G. F. Uhlenbeck and S. Goudsmit, *Naturwissenschaften*, **13**, 953 (1925).

Table 7.1

Electronic Configuration of the Elements

Atomic No.	Element	1 s	2 s	2 p	3 s	3 p	3 d	4 s	4 p	4 d	5 s	5 p
1	H	1										
2	He	2										
3	Li	2	1									
4	Be	2	2									
5	B	2	2	1								
6	C	2	2	2								
7	N	2	2	3								
8	O	2	2	4								
9	F	2	2	5								
10	Ne	2	2	6								
11	Na	2	2	6	1							
12	Mg	2	2	6	2							
13	Al	2	2	6	2	1						
14	Si	2	2	6	2	2						
15	P	2	2	6	2	3						
16	S	2	2	6	2	4						
17	Cl	2	2	6	2	5						
18	Ar	2	2	6	2	6						
19	K	2	2	6	2	6		1				
20	Ca	2	2	6	2	6		2				
21	Sc	2	2	6	2	6	1	2				
22	Ti	2	2	6	2	6	2	2				
23	V	2	2	6	2	6	3	2				
24	Cr	2	2	6	2	6	5	1				
25	Ma	2	2	6	2	6	5	2				
26	Fe	2	2	6	2	6	6	2				
27	Co	2	2	6	2	6	7	2				
28	Ni	2	2	6	2	6	8	2				
29	Cu	2	2	6	2	6	10	1				
30	Zn	2	2	6	2	6	10	2				
31	Ga	2	2	6	2	6	10	2	1			
32	Ge	2	2	6	2	6	10	2	2			
33	As	2	2	6	2	6	10	2	3			
34	Se	2	2	6	2	6	10	2	4			
35	Br	2	2	6	2	6	10	2	5			
36	Kr	2	2	6	2	6	10	2	6			
37	Rb	2	2	6	2	6	10	2	6		1	
38	Sr	2	2	6	2	6	10	2	6		2	
39	Y	2	2	6	2	6	10	2	6	1	2	
40	Zr	2	2	6	2	6	10	2	6	2	2	
41	Cb	2	2	6	2	6	10	2	6	4	1	
42	Mo	2	2	6	2	6	10	2	6	5	1	
43	Tc	2	2	6	2	6	10	2	6	6	1	
44	Ru	2	2	6	2	6	10	2	6	7	1	
45	Rh	2	2	6	2	6	10	2	6	8	1	
46	Pd	2	2	6	2	6	10	2	6	10		
47	Ag	2	2	6	2	6	10	2	6	10	1	
48	Cd	2	2	6	2	6	10	2	6	10	2	
49	In	2	2	6	2	6	10	2	6	10	2	1
50	Sn	2	2	6	2	6	10	2	6	10	2	2
51	Sb	2	2	6	2	6	10	2	6	10	2	3
52	Te	2	2	6	2	6	10	2	6	10	2	4
53	I	2	2	6	2	6	10	2	6	10	2	5

Atomic No.	Element	1 s	2 s	2 p	3 s	3 p	3 d	4 s	4 p	4 d	4 f	5 s	5 p	5 d	5 f	6 s	6 p	6 d	7 s
54	Xe	2	2	6	2	6	10	2	6	10		2	6						
55	Cs	2	2	6	2	6	10	2	6	10		2	6			1			
56	Ba	2	2	6	2	6	10	2	6	10		2	6			2			
57	La	2	2	6	2	6	10	2	6	10		2	6	1		2			
58	Ce	2	2	6	2	6	10	2	6	10	2	2	6			2			
59	Pr	2	2	6	2	6	10	2	6	10	3	2	6			2			
60	Nd	2	2	6	2	6	10	2	6	10	4	2	6			2			
61	Pm	2	2	6	2	6	10	2	6	10	5	2	6			2			
62	Sm	2	2	6	2	6	10	2	6	10	6	2	6			2			
63	Eu	2	2	6	2	6	10	2	6	10	7	2	6			2			
64	Gd	2	2	6	2	6	10	2	6	10	7	2	6	1		2			
65	Tb	2	2	6	2	6	10	2	6	10	9	2	6			2			
66	Dy	2	2	6	2	6	10	2	6	10	10	2	6			2			
67	Ho	2	2	6	2	6	10	2	6	10	11	2	6			2			
68	Er	2	2	6	2	6	10	2	6	10	12	2	6			2			
69	Tm	2	2	6	2	6	10	2	6	10	13	2	6			2			
70	Yb	2	2	6	2	6	10	2	6	10	14	2	6			2			
71	Lu	2	2	6	2	6	10	2	6	10	14	2	6	1		2			
72	Hf	2	2	6	2	6	10	2	6	10	14	2	6	2		2			
73	Ta	2	2	6	2	6	10	2	6	10	14	2	6	3		2			
74	W	2	2	6	2	6	10	2	6	10	14	2	6	4		2			
75	Re	2	2	6	2	6	10	2	6	10	14	2	6	5		2			
76	Os	2	2	6	2	6	10	2	6	10	14	2	6	6		2			
77	Ir	2	2	6	2	6	10	2	6	10	14	2	6	9		0			
78	Pt	2	2	6	2	6	10	2	6	10	14	2	6	9		1			
79	Au	2	2	6	2	6	10	2	6	10	14	2	6	10		1			
80	Hg	2	2	6	2	6	10	2	6	10	14	2	6	10		2			
81	Tl	2	2	6	2	6	10	2	6	10	14	2	6	10		2	1		
82	Pb	2	2	6	2	6	10	2	6	10	14	2	6	10		2	2		
83	Bi	2	2	6	2	6	10	2	6	10	14	2	6	10		2	3		
84	Po	2	2	6	2	6	10	2	6	10	14	2	6	10		2	4		
85	At	2	2	6	2	6	10	2	6	10	14	2	6	10		2	5		
86	Rn	2	2	6	2	6	10	2	6	10	14	2	6	10		2	6		
87	Fr	2	2	6	2	6	10	2	6	10	14	2	6	10		2	6		1
88	Ra	2	2	6	2	6	10	2	6	10	14	2	6	10		2	6		2
89	Ac	2	2	6	2	6	10	2	6	10	14	2	6	10		2	6	1	2
90	Th	2	2	6	2	6	10	2	6	10	14	2	6	10		2	6	2	2
91	Pa	2	2	6	2	6	10	2	6	10	14	2	6	10	2	2	6	1	2
92	U	2	2	6	2	6	10	2	6	10	14	2	6	10	3	2	6	1	2
93	Np	2	2	6	2	6	10	2	6	10	14	2	6	10	4	2	6	1	2
94	Pu	2	2	6	2	6	10	2	6	10	14	2	6	10	6	2	6		2
95	Am	2	2	6	2	6	10	2	6	10	14	2	6	10	7	2	6		2
96	Cm	2	2	6	2	6	10	2	6	10	14	2	6	10	7	2	6	1	2
97	Bk	2	2	6	2	6	10	2	6	10	14	2	6	10	8	2	6	1	2
98	Cf	2	2	6	2	6	10	2	6	10	14	2	6	10	10	2	6		2
99	Es	2	2	6	2	6	10	2	6	10	14	2	6	10	11	2	6		2
100	Fm	2	2	6	2	6	10	2	6	10	14	2	6	10	12	2	6		2
101	Md	2	2	6	2	6	10	2	6	10	14	2	6	10	13	2	6		2

We mention here that the Pauli principle applies not only to atomic electrons but to any system containing several identical particles of half-integer spin (we shall see presently what that means). Such particles are called fermions and whenever several identical fermions (electrons, protons, or neutrons are some examples) get close together, we can be assured that no two of them will have the same set of quantum numbers. There are other particles called "bosons" (π-mesons and photons are examples) that do not mind so much and that will happily wear a set of quantum numbers worn by another boson of the same kind at the same place.

7.3 SPIN

Part of the angular momentum of the hydrogen atom results from the motion of the electron around the proton. This part of the angular momentum, the one we discussed in Chapter 5.8-9, is, therefore, often called the **orbital angular momentum.** Besides the orbital angular momentum that all particles can have, most elementary particles have an **intrinsic angular momentum.** Regardless of their state of motion, they behave somewhat like little gyros or tops. We call this intrinsic angular momentum the **spin.** Particles with spin are electrons, protons, neutrons, and photons, to name a few. Particles without spin are π-mesons and K-mesons.

The spin of a particle cannot be explained in terms of the non-relativistic quantum mechanics we have developed here. To understand it, we would have to find the eigenfunctions of a relativistically invariant Hamiltonian, first proposed by Dirac. This would go considerably beyond the scope of this book, and we introduce spin here as an empirically known property of some elementary particles. We saw in Chapter 5.9 that in a hydrogen atom the orbital angular momentum is characterized by a quantum number l, and that the expectation value of the magnitude of the angular momentum is

$$\sqrt{l(l+1)}\,\hbar$$

where

$$l = 0, 1, 2, \ldots, n-1$$

The expectation value of the z-component of the angular momentum is $m\hbar$ where m is an integer

$$-l \leqslant m \leqslant l$$

The spin angular momentum can be treated in a similar fashion. The spin quantum number that is the equivalent of l is usually called I; it describes the magnitude

$$\sqrt{I(I+1)}\,\hbar$$

of the spin angular momentum. The quantum number m_I, the equivalent of m (or m_l as we shall write it from now on to distinguish the two) can again have values

$$-I \leqslant m_I \leqslant I$$

changing in steps of one. m_I describes the z-component

$$m_I \hbar$$

of the spin angular momentum. There is, however, one *significant* and surprising *difference*: **The spin quantum numbers are capable of half-integer values.** Table 7.2 lists the possible values of I and m_I up to $I = \frac{3}{2}$. Electrons have

Table 7.2

I	0	$\frac{1}{2}$	1	$\frac{3}{2}$
m_I	0	$-\frac{1}{2}, \frac{1}{2}$	$-1, 0, 1$	$-\frac{3}{2}, -\frac{1}{2}, \frac{1}{2}, \frac{3}{2}$

$I = \frac{1}{2}$. This explains the fact that in the periodic table (Table 7.1) the possible sets of the orbital quantum numbers are represented just twice. The states with the same values of n, l, and m can still differ in m_I, which can be either $+\frac{1}{2}$ or $-\frac{1}{2}$. Protons and neutrons also have $I = \frac{1}{2}$. Some of the esoteric particles manufactured with the large high-energy accelerators have higher values of I. Particles with half-integer spin

$$I = \frac{1}{2}, \frac{3}{2}, \text{etc.}$$

are called **fermions** and obey the Pauli principle. Particles with integer spin

$$I = 0, 1, 2, 3, \text{etc.}$$

are called **bosons** and could not care less.

7.4 THE MAGNETIC MOMENT OF THE ELECTRON

If a particle moves in a central field in such a way that its orbit encloses an area, its angular momentum is proportional to this area. If the particle is charged, its orbital motion constitutes a current flowing around the periphery of the area, resulting in a magnetic moment. We would expect that these statements of classical physics remain true in quantum mechanics and that an electron in a state with $l \neq 0$ has an orbital magnetic moment. This is, indeed, true. If we solve the Schrödinger equation of an atom in a magnetic field **B** including the terms for the magnetic part of the total energy, we find that the energy eigenvalues contain terms of the form $\boldsymbol{\mu} \cdot \mathbf{B}$ where $\boldsymbol{\mu}$ is a magnetic dipole moment. These calculations are not trivial, and we shall forego them here, substituting some semiclassical arguments leading to *exactly* the same value for μ.

Let an electron move in a circular orbit of radius $r = \hbar^2/me^2$ around a proton. Let us assume that the z-component of the angular momentum is $L_z = \hbar$. We equate this with the classical angular momentum:

$$L_z = \hbar = m_e r^2 \omega \qquad (7.9)$$

An electron orbiting the proton with a frequency $\omega/2\pi = \nu = 1/\tau$ constitutes a current of $i = \nu$ electrons/sec. Such a current enclosing an area $A = \pi r^2$ produces a magnetic dipole moment:

$$\mu_0 = iA = \frac{e\omega\pi r^2}{2\pi} \qquad (7.10)$$

Substitution of Eq. 7.9 into Eq. 7.10 yields

$$\mu_0 = \frac{e\hbar}{2m_e} \qquad (7.11)$$

This is the so-called **Bohr magneton.** An electron in a state with a total angular momentum $\sqrt{l(l+1)}\,\hbar$ has a magnetic moment

$$\mu = \sqrt{l(l+1)}\mu_0 \qquad (7.12)$$

If the z-component of the angular momentum is $m_l\hbar$ the z-component of the magnetic moment will be

$$\mu_l = m_l\mu_0 \qquad (7.13)$$

All this is borne out by a rigorous quantum mechanical calculation and verified by experiments.

The intrinsic angular momentum of the electron is *also* connected with a magnetic moment, the so-called intrinsic or **spin magnetic moment.** Surprisingly this magnetic moment is the *same* as the orbital magnetic moment belonging to $m_l\hbar = \hbar$ although the *spin angular momentum* $m_l\hbar$ is only half as big. This is often expressed by saying that the **gyromagnetic ratio** of the electron (also called its g-value) is equal to two.

$$g = \frac{\text{spin magnetic moment measured in Bohr magnetons}}{\text{angular momentum measured in units of } \hbar} \qquad (7.14)$$

This result is correctly given by Dirac's relativistic quantum mechanics, and particles with $g = 2$ are often called **Dirac particles.** Obviously Dirac particles are fermions. The only known Dirac particles are the electron, and the muon, and their antiparticles. A very precise measurement of g shows a small deviation from Dirac's value. The g-values of the electron and muon are

$$g_e = 2.003192 \qquad (7.15)$$

and

$$g_\mu = 2.00233 \qquad (7.16)$$

This very small deviation from $g = 2$ and even the small difference between g_e and g_μ are accounted for by a more complete relativistic quantum theory called *quantum electrodynamics*.

7.5 THE VECTOR MODEL

Now we extend our semi-quantitative contemplation of many electron atoms to their angular momenta. We do this for three reasons:

1. This approach gives many of the essential features of atomic structure with much less bother than a complete quantum-mechanical treatment of the subject.
2. It will enable us to understand the nomenclature used to describe atomic states.
3. It demonstrates—on firm ground—a technique often used at the frontiers of theoretical physics: one picks an experimentally observed or theoretically expected symmetry, here rotational symmetry, and derives all the properties related to it, even though one lacks a complete understanding of the system (here the many-electron atom).

The most surprising aspect of the hydrogen eigenfunctions is, certainly, that even the component of the angular momentum vector with regard to an arbitrary direction is capable only of certain discrete values.

This was confirmed in a famous experiment by Stern and Gerlach[4] and it is true for all quantum-mechanical systems. There are no special requirements for establishing the direction with respect to which an angular momentum can have only integer or half-integer components. Therefore, we define a direction like this:

Fig. 7.1   This is *one* way to establish an axis of quantization.

---

[4] W. Gerlach and O. Stern, *Z. f. Phys.*, **7**, 249 (1921). For a review of this important experiment, see R. M. Eisberg, *Fundamentals of Modern Physics*, John Wiley and Sons, New York, 1961.

forcing every electron in the universe to adjust its angular momentum components in the approved manner. What is the catch? Well, it is this: Our thumb, of course, does not lift any existing degeneracy, and an electron with $l = 3$ can have eigenfunctions with $m_l = -3, -2, -1, 0, 1, 2,$ or 3 but it can *also* have any linear combination of these eigenfunctions. Thus what we have really said is this: An electron of angular momentum l can be described in *any* coordinate system by a superposition of at most $2l + 1$ eigenfunctions with m_l ranging from $-l$ to $+l$. This is why, so far, the quantum number m_l has not taken on much meaning for us. It has not led to observable consequences. In a strictly central potential, all $2l + 1$ states of different m_l but equal l are degenerate. They have the same energy and cannot be distinguished spectroscopically. This is changed if we introduce terms into the Hamiltonian that depend on the angle ϑ. Such terms can be introduced through an external electric or magnetic field. It is obvious that in the presence of such fields the energy will depend on the orientation of the angular momentum, and thereby the magnetic moment, in the external field.

Let us assume that a hydrogen atom is in a state in which it has a total orbital angular momentum $\sqrt{l(l + 1)}\, \hbar$. If we place the atom into a magnetic field in the z direction, the field will exert a torque on the magnetic dipole moment, trying to align it with the z-direction. Since the atom has an angular momentum it will behave like any gyroscope in such a situation. The torque will not align it but will cause the axis of the angular momentum to precess around the z direction. This is shown in Figure 7.2. The precession results in a constant value of the z component of the angular momentum, and this constant value *has to be* one of the allowed values $m_l \hbar$ where $-l \leqslant m_l \leqslant l$. The x and y component of the angular momentum keep changing all the time and, therefore, cannot be measured simultaneously. This is obviously what the commutation relations (Eqs. 6.101 and 6.102) have been trying to tell us all along.

Which value of m_l the atom will assume depends on the orientation it had when it was first brought into the field. Actually, we do not even need an external field to define a special direction for an atom. Any electron in a many-electron atom has either a spin or a spin and an orbital angular momentum. Any electron can thus define a direction with respect to which the other electrons must adjust their angular momentum components.

Now we investigate what happens when the magnetic moments of two electrons react with each other. As an example we consider a p-electron and a d-electron, disregarding for the moment that the two electrons also have a spin angular momentum. The magnetic moments of both electrons will try to align each other and, in the process, will set up a precession around a common axis. The resultant of the two angular momenta in the direction of this axis has to be quantized to have a magnitude that can be expressed as

$$\sqrt{L(L + 1)}\, \hbar$$

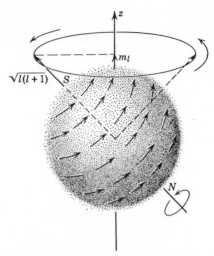

Fig. 7.2 A torque acting on an atom with angular momentum (or on any gyroscope) will not align the axis of the angular momentum but will cause it to precess. The projection of the angular momentum on the axis of precession is an integer multiple of $\hbar$. For a macroscopic gyroscope the number of possible orientations is so large as to be continuous.

where L is an integer. For the example given, there exist only three ways in which this can be accomplished. The three possibilities are shown in Figure 7.3a to 7.3c. Actually this figure makes the situation look more complicated than it really is. The possible values of L in our example are simply

$$L = 2 + 1, \quad L = 2 \pm 0, \quad \text{and} \quad L = 2 - 1 \qquad (7.17)$$

In other words, L is the sum of the larger of the two l's, and the possible values m_l of the smaller of the two. For two arbitrary angular momentum quantum numbers l and l' with $l \geqslant l'$, we thus obtain

$$l - l' \leqslant L \leqslant l + l' \qquad (7.18)$$

In the most general case of the addition of n orbital angular momenta, the highest possible value of L is given by

$$L_{\max} = \sum_1^n l_k \qquad (7.19)$$

where l_k is the quantum number of the orbital angular momentum of the kth electron. The lowest value of L is zero or the lowest nonnegative number that we obtain if we subtract all the other l_k from the largest l_k.

Spin angular momenta add in a similar fashion. If there are two electrons, the possible values of the total spin S are[5]

$$S = \tfrac{1}{2} + \tfrac{1}{2} = 1, \quad \text{and} \quad S = \tfrac{1}{2} - \tfrac{1}{2} = 0$$

[5] It is customary to use the letter I for the spin of elementary particles and nuclei, whereas the letter S is used for the total spin of many electron atoms.

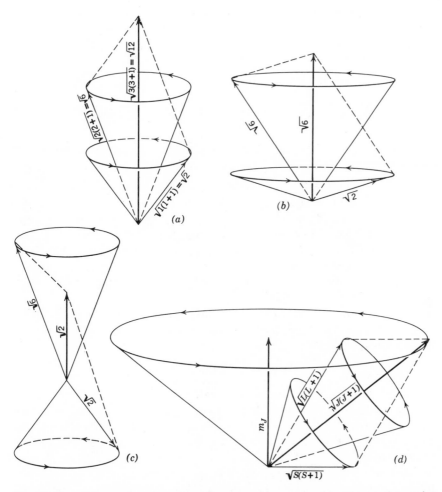

Fig. 7.3*a,b,c* Two angular momenta precess around a common axis. The opening angles of the precession cones adjust themselves so that the resultant has a length that can be expressed as $\sqrt{L(L + 1)}\, \hbar$, where L is an integer.

Fig. 7.3*d* The total angular momentum described by the quantum number J will align itself in such a way that its component with respect to a given external axis has one of the possible values of m_J.

For three electrons, we get

$$S = \tfrac{1}{2} + \tfrac{1}{2} + \tfrac{1}{2} = \tfrac{3}{2}, \quad \text{and} \quad S = \tfrac{1}{2} + \tfrac{1}{2} - \tfrac{1}{2} = \tfrac{1}{2}$$

For n electrons, S can have any value between

$$S = 0 \text{ and } S = n/2 \quad \text{or} \quad S = \tfrac{1}{2} \text{ and } S = n/2$$

depending on whether n is even or odd. It is obvious that in a many-electron atom, where every electron has a spin and many have an orbital angular momentum, things could be appallingly complicated. Fortunately, two factors cooperate to simplify matters.

1. The Pauli principle sees to it that the angular momentum of a closed shell is zero. This leaves us with only the outermost electrons to worry about.

2. Spins and orbital angular momenta (at least, in the light atoms) obey Russel Saunders coupling also called L-S coupling.

These terms mean the following. In a many-electron atom all the orbital angular momenta add up to a total orbital angular momentum $\sqrt{L(L + 1)}\,\hbar$ in the manner just discussed. All the electron spins add up to a total spin angular momentum $\sqrt{S(S + 1)}\,\hbar$ as discussed above.

The total spin and the total orbital angular momentum align themselves with respect to each other in such a way that they add up (vector fashion) to a total angular momentum characterized by a quantum number J. J can assume any value between $J = |L - S|$ and $J = |L + S|$, varying in steps of one. Thus, J is either always an integer or always a half-integer depending on the nature of S.

To illustrate this we return to the example of the p and d electron (Figure 7.3a to 7.3c). Let us assume that the two electrons have arranged themselves in such a way that

$$L = 3 \qquad \text{and} \qquad S = 1$$

The *total orbital angular momentum* and the *total spin* will now precess around each other and form a resultant total angular momentum (see Figure 7.3d), characterized by the new quantum number J, which can be either 2, 3, or 4. If the two electrons had added up to $L = 2$, $S = 1$, J could have been 1, 2, or 3, etc. Now comes the important point: If the atom is brought into a magnetic field, which is weak compared with the extremely strong magnetic fields inside the atom, it will precess as a whole and adjust its total angular momentum in such a way that its z-component

$$m_J\hbar$$

has one of the values:

$$-J\hbar, -(J - 1)\hbar, \ldots, (J - 1)\hbar, J\hbar$$

This latter $2J + 1$-fold degeneracy which is lifted only by an external magnetic field, is called the multiplicity of the state.

Notice that if the total angular momentum aligns itself in a magnetic field in such a way that m_J is an integer or half integer, neither m_s, m_l, m_S, or m_L are integers or half-integers. Actually these z-components are not even constant but vary in time as a result of the precession around the common axis. Therefore, they cannot be used to describe the state of the system.

This is often expressed by saying that m_l, m_s, m_L and m_S are *not* "**good quantum numbers**" if several angular momenta couple with each other. If the external field becomes stronger than the internal fields, it is "every spin for itself" and m_l and m_s become good quantum numbers. In other words, the spin and orbital angular momentum of each individual electron adjusts itself so that its components with regard to the external magnetic field become integers or, in the case of spin, half-integers. In such a case we often say, colloquially, that the strong magnetic field *decouples* spin and orbital angular momenta.

Based on this coupling scheme is the following notation. The total orbital angular momentum is expressed by a capital letter:

$$S, P, D, F, \text{etc.}$$

corresponding to

$$L = 0, 1, 2, 3, \text{etc.}$$

The quantum number of the total angular momentum J is tacked on as a suffix, and the number of possible orientations of the total spin $2S + 1$ is expressed by a prefix. Thus the system we have discussed would be in a

$$^3F_2, \qquad ^3F_3, \qquad \text{or} \qquad ^3F_4 \text{ state}$$

depending on the actual value of J. If the principal quantum number is of interest, it can be added in front of the letter. To gain more familiarity with this notation we apply it to some concrete examples.

Hydrogen

There is only one electron, hence $l = L$ and $S = \frac{1}{2}$. The ground state is obviously a $1\,^2S_{1/2}$ (pronounced: doublet S one-half) state.

In the first excited state, L can be either 0, in which case $J = \frac{1}{2}$, leading to a

$$2\,^2S_{1/2} \text{ state,}$$

or it can be 1. In this case the spin has two possible orientations leading to $J = \frac{1}{2}$ or $J = \frac{3}{2}$ or, in our new notation, to a

$$2\,^2P_{1/2} \text{ state} \qquad \text{or} \qquad \text{a } 2\,^2P_3 \text{ state}$$

These two states have different energies since in one case the magnetic moments of spin and orbital angular momentum are antiparallel and in the other case they are parallel with respect to each other. This results in different values of the magnetic energy. Since the magnetic energy is small compared to the Coulomb energy, this energy difference is very small. It leads to a splitting of the spectral lines and is called **fine structure** splitting.

We see now why $2S + 1$ and not S is used as a prefix. $2S + 1$ is the observed fine structure splitting that results from the interaction of spin and orbital angular momentum if $L \geqslant S$.[6]

The $2\,{}^2S_{1/2}$ and $2\,{}^2P_{1/2}$ states remain degenerate under the influence of the *spin orbit interaction*. Very precise experiments, first performed by W. Lamb,[7] showed that even this degeneracy is lifted by an interaction with the radiation field. This **Lamb shift** between the $2\,{}^2S_{1/2}$ and the $2\,{}^2P_{1/2}$ state is only about $\frac{1}{10}$ of the already very small fine structure splitting.

Helium

Because of the Pauli principle, the ground state of He must be a

$$1\,{}^1S_0 \text{ state}$$

The first excited state can be an S state or a P state. Let us first consider the S state. Since the two electrons now have different principal quantum numbers the Pauli principle no longer enforces opposite spin directions. Hence, we can have either a

$$2\,{}^1S_0 \text{ state} \qquad \text{or} \qquad \text{a } 2\,{}^3S_1 \text{ state}$$

$$\text{(singlet } s \text{ zero)} \qquad \text{(triplet } s \text{ one)}$$

Now to the P state. Here S can be either zero (coupling with $L = 1$ to give $J = 1$) or it can be one (coupling with $L = 1$ to give $J = 0, 1,$ or 2). Thus the resulting states are

$$2\,{}^1P_1, 2\,{}^3P_0, 2\,{}^3P_1, \qquad \text{and} \qquad 2\,{}^3P_2$$

It is interesting to note that a *selection rule*

$$\Delta S = 0$$

allows transitions between two states only if the initial and final states have the same total spin. Transitions between triplet and singlet states are thus forbidden. This results (in the case of He) in the existence of two independent series of spectral lines where no line in the *singlet spectrum* can be expressed as a difference of energy levels participating in the *triplet spectrum*, and vice versa. Before this was properly understood there were thought to be two kinds of helium, *ortho-helium* and *para-helium*, chemically alike but, somehow, spectroscopically different.

[6] For $L < S$ the level splits into $2L + 1$ different levels.
[7] W. Lamb and R. C. Retherford, *Phys. Rev.*, **79**, 549 (1950).

Problems

7.1   Estimate the ionization energy of the carbon atom.

7.2   Why is it difficult to separate the rare earths chemically?

7.3   Which of the following elements: Li, Be, N, F, Na, Mg, Cl, would you expect to have a spectrum similar to that of hydrogen? Which would have a spectrum similar to that of helium?

7.4   The He nucleus has no spin. Draw a qualitative diagram of *all* the energy levels with $n = 1$, $n = 2$, $n = 3$ of the singly ionized He atom (a) without an external magnetic field, and (b) with an external magnetic magnetic field.

7.5   Draw a schematic diagram of levels with $n = 1$, $n = 2$, $n = 3$, of the muonic carbon atom. (Consider only the energy levels resulting from the various states of the muon and assume that the electrons remain in their ground states.)

7.6   Which of the following states cannot exist: $^2P_{1/2}$, $^2P_1$, $^3P_{1/2}$, $^2P_{5/2}$, $^2P_{1/2}$, $^1P_0$, $^3P_0$, $^3P_1$?

7.7   How many different energy levels does a $^4P_{1/2}$, and a $^2P_{3/2}$ state have (a) without an external magnetic field, and (b) in an external magnetic field?

7.8   Neutrons and protons are fermions with spin $I = \frac{1}{2}$. What total nuclear spin would you expect the following nuclei to have: H^2, H^3, He^3, He^4?

7.9   Pions have no spin. A negative pion can be captured into atomic states by a proton, forming a pionic hydrogen atom. The proton has spin $\frac{1}{2}$. Draw a schematic energy level diagram for the lowest s, p, and d state of a pionic hydrogen atom (a) in the absence of a magnetic field, and (b) in a magnetic field.

7.10   The electron and its positive antiparticle the positron both have spin $\frac{1}{2}$. Since they have opposite charges they can form a hydrogenlike atom that is called positronium. The ground state of positronium is, of course, a $1s$ state. Would you expect, in the absence of an external magnetic field, that the ground state is split as a result of the interaction between electron and positron spin? If so, into how many levels?

7.11   The hydrogen molecule H_2 consists of two hydrogen atoms. These atoms are bound together by the two electrons that orbit around *both* nuclei. The wave function of both these electrons is spherically symmetrical (s-state). Yet, there exist two, experimentally discernible varieties (states) of hydrogen at low temperatures, *ortho*-hydrogen and *para*-hydrogen. What distinguishes these two kinds of hydrogen?

7.12   (a) Given three particles with angular momentum quantum numbers $l_1 = 4$, $l_2 = 3$, $l_4 = 2$, in how many different ways can these angular momenta be combined to give a total angular momentum L? (The values of L obtained in this way are not necessarily different from each other).
(b) Into how many states (not necessarily all of different energy) do the states

obtained in (a) split under the influence of an external magnetic field? Show that this latter number is the same whether the angular momenta interact with each other, or not.

SOLUTIONS

7.7 In a $^4P_{1/2}$ state the angular momentum quantum numbers have the following values: $L = 1$ (P state), and $S = \frac{3}{2}$ (prefix 4 = $2S + 1$). The two have combined to give $J = \frac{1}{2}$ (suffix $\frac{1}{2}$). The $^4P_{1/2}$ state is one of the three fine structure levels, each belonging to one of the three possible values of the total angular-momentum quantum number $J = \frac{5}{2}$, $J = \frac{3}{2}$, and $J = \frac{1}{2}$. As such it is represented by a single-energy level in the absence of a magnetic field. In an external magnetic field (or, also, because of an interaction with a nuclear spin) the total angular momentum described by J (here $J = \frac{1}{2}$) can orient itself in two ways so that $m_J = \frac{1}{2}$ or $m_J = -\frac{1}{2}$, resulting in a split into two different levels. In the case of the $^2P_{3/2}$ state the quantum numbers are $L = 1$, $S = \frac{1}{2}$, and $J = \frac{3}{2}$. This state is also represented by a single-energy level in the absence of a magnetic field. The level is one of the two fine structure levels belonging to $J = \frac{3}{2}$ and $J = \frac{1}{2}$. In the presence of a magnetic field there will be four levels belonging to $m_J = \frac{3}{2}$, $m_J = \frac{1}{2}$, $m_J = -\frac{1}{2}$, and $m_J = -\frac{3}{2}$.

7.12 We start with the particle with $l_1 = 4$. Relative to this particle the particle with $l_2 = 3$ can orient itself in such a manner that the resulting angular momentum has one of the quantum numbers $L' = 7$, 6, 5, 4, 3, 2, or 1. Relative to these possible angular momenta, the remaining particle can orient itself so that m_{l_3} is either 2, 1, 0, -1, or -2. Hence,

m_{l_3}	$L' = 7$	6	5	4	3	2	1
2	9	8	7	6	5	4	3
1	8	7	6	5	4	3	2
0	7	6	5	4	3	2	1
−1	6	5	4	3	2	1	
−2	5	4	3	2	1	0	

(Note that in the last column l_3 is the larger of the two numbers so that it is now L' that has to orient itself so that $m_{L'} = \pm 1$, or 0.)

We have thus obtained a total of 33 different combinations. We would, of course, have obtained the same number if, for example, we had started by combining l_2 and l_3 and the result L'' with l_1. In the presence of a magnetic field, $L = 9$ splits into $2L + 1 = 19$ different levels and $L = 8$ into $2L + 1 = 17$, etc. Adding it all up, we obtain $19 + 2 \cdot 17 + \cdots = 315$. So much for the interacting particles. If there is no interaction, the first particle can assume any of $2l_1 + 1 = 9$ different orientations. Independently, the second particle can assume $2l_2 + 1 = 7$ and the third $2l_3 + 1 = 5$. The total number of different combinations is obviously $9 \cdot 7 \cdot 5 = 315$, as it should be.

8

LIGHT EMISSION

8.1 TRANSITION PROBABILITY

It is well known that an atom in an excited state, described by a wave function $u_{n'J'm_{J'}}$ can return spontaneously to the ground state u_{nJm_J}, emitting a photon whose energy $\hbar\omega$ equals the energy difference between the initial and the final state of the atom. A complete quantum mechanical theory of light emission should be able to predict the probability for such a transition. Since the final state of the *system* contains a photon and the initial state does not, a theory that describes light emission has to be able to describe the creation of a photon. Such a theory exists and is known as quantum electrodynamics. Regrettably, quantum electrodynamics, which is a branch of quantum field theory, goes far beyond the scope of this book.

To stay out of the mathematical brierpatch of quantum field theory we shall use the technique that has served us well thus far: We shall write down the expression for the analogous classical situation and then translate it into quantum mechanics. Since photons do not naturally emerge from this theory we must introduce them as *deus ex machina* at the appropriate time.

A classical device that radiates light, although of a long wavelength, is an antenna or electric dipole. The flow of electromagnetic energy, according to classical electrodynamics, is given by the time average of the Poynting vector

$$S = [EH] \tag{8.1}$$

where E and H are the vectors of the electric and magnetic field. The Poynting vector of the radiation field of an electric dipole, radiating with a frequency ω, is[1]

$$|S| = \frac{(ed)^2\omega^4 \sin^2\vartheta \sin^2(r/\lambdabar - \omega t)}{4\pi c^3 r^2} \; ; \quad \left(\lambdabar = \frac{\lambda}{2\pi}\right) \tag{8.2}$$

where ed is the electric dipole moment (see Figure 8.1).

[1] See Marion, *Classical Electromagnetic Radiation*, Academic Press, New York and London 1965.

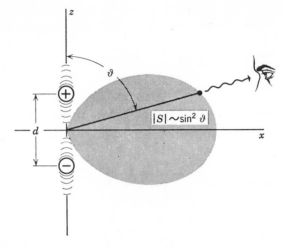

Fig. 8.1 According to Maxwell's theory the intensity of the light emitted by an electric dipole is proportional to $\sin^2 \vartheta$.

The radiation of a classical dipole is thus proportional to the square of its (oscillating) dipole moment. We, therefore, investigate the quantum mechanical expectation value of the dipole moment $e\mathbf{d}$ of a stationary state characterized by an eigenfunction $u(r)$.

$$\langle e\mathbf{d} \rangle = \int u^*e\mathbf{d}\, u\, d\tau = e \int \mathbf{d}u^*u\, d\tau \tag{8.3}$$

If we change $\mathbf{r}$ into $-\mathbf{r}$, u^*u remains unchanged if u has a definite parity; however, because of $\mathbf{d}$, the integrand changes sign. Since any integral of the type

$$\int f(x, y, z)\, d\tau$$

vanishes between symmetrical limits if[2]

$$f(x, y, z) = -f(-x, -y, -z) \tag{8.4}$$

we can pronounce the following **theorem:**

The stationary states of a system whose Hamiltonian is symmetrical about the origin have no electric dipole moment.

Does this demolish our hopes of explaining the emission of light by atoms? Of course not. Stationary states do not radiate anyway. What we are after

---

[2] The proof of this statement is straightforward in one dimension and can easily be extended to three dimensions.

must be a transient oscillating dipole moment occurring when an atom goes from one stationary state to another. To investigate this possibility we return to the time-dependent Schrödinger equation (Eq. 2.15):

$$i\hbar \frac{\partial \psi}{\partial t} = -\frac{\hbar^2}{2m} \nabla^2 \psi + V(\mathbf{r})\psi \tag{2.15}$$

According to Eq. 2.21, this equation has the solution:

$$\psi = e^{-i\omega t}u(\mathbf{r}) = e^{-(iEt/\hbar)}u(\mathbf{r}) \tag{8.5}$$

If there is more than one stationary state (and this is the only case of interest to us) all the linear combinations

$$\psi = \sum_{k=1} c_k e^{-(iE_k t/\hbar)}u_k(\mathbf{r}) \tag{8.6}$$

are solutions of Eq. 2.15. The symbol k stands here as an abbreviation for a set of quantum numbers n, J, m_J. Thus the most general—nonstationary—probability density is, according to Eq. 8.6:

$$\psi^*\psi = \sum_{i=1}\sum_{k=1} c_i^* c_k \exp\left[\frac{i(E_i - E_k)t}{\hbar}\right]u_i^*(\mathbf{r})u_k(\mathbf{r}) \tag{8.7}$$

The probability density (Eq. 8.7) fluctuates in time with the frequencies

$$\omega_{ik} = -\frac{E_i - E_k}{\hbar} \tag{8.8}$$

This leads us to suspect that light of these frequencies might be emitted *if* these fluctuations are connected with a changing electric dipole moment of the atom. Substitution of Eq. 8.7 into Eq. 8.3 yields the expectation value of the electric dipole moment:

$$\langle e\mathbf{d}\rangle = e\sum_{i=1}\sum_{k=1} c_i^* c_k e^{-i\omega_{ik}t}\int d u_i^* u_k \, d\tau \tag{8.9}$$

This is the most general form of the electric dipole moment. A transition from an initial state k to a final state i resulting in the emission of a photon of frequency ω_{ik} is represented in Eq. 8.9 by the term:

$$\langle e\mathbf{d}\rangle_{ik} = ec_i^* c_k e^{-i\omega_{ik}t}\int d u_i^* u_k \, d\tau \tag{8.10}$$

Now we make the decisive *assumption*: The intensity of the light emitted by an atom is given by the time average of Eq. 8.2 *if* we replace $(e\mathbf{d})^2$ with $\langle 2e\mathbf{d}\rangle_{ik}^2$. The factor of two results from the fact that the exchange of i and k in Eq. 8.10 leaves the time average unchanged. Thus the contribution with the frequency $|\omega_{ik}|$ appears twice in the sum (Eq. 8.9.) The time average of $\sin^2 (r/\lambda - \omega t)$ is

$$\langle \sin^2 (r/\lambda - \omega t) \rangle = \tfrac{1}{2} \tag{8.11}$$

so that the light intensity becomes, according to Eq. 8.2,

$$I_{ik} = \frac{\langle ed \rangle_{ik}^2 \omega_{ik}^4 \sin^2 \vartheta}{2\pi c^3 r^2} \tag{8.12}$$

If we integrate Eq. 8.12 over the surface of a sphere of radius $r = 1$, we obtain the luminous intensity of the light source, i.e., the energy emitted by the atom per unit time:

$$L_{ik} = \int I_{ik} r^2 \sin \vartheta \, d\vartheta \, d\varphi = \frac{4\omega_{ik}^4}{3c^3} \langle ed \rangle_{ik}^2 \tag{8.13}$$

Obviously, it is not proper to speak of light intensity if we are dealing with the light emitted by a single atom. An atom emits individual photons, and a quantum-mechanical theory should tell us the probability of finding them. Here the shortcomings of our theory become apparent. Our method of substituting the expectation value of the electric dipole moment into the classical formula (Eq. 8.2) simply does not describe the creation of photons but rather the emission of a continuous wave. However, all is not lost. Instead of carrying out a *quantization of the radiation field* (as we properly should[3]), we simply introduce photons, ad hoc, into our theory.

Now, if L is the luminous intensity of a source, the number of photons emitted by the source-per-unit time must be given by

$$N = \frac{L}{\hbar \omega} \tag{8.14}$$

If we have only one atom and L_{ik} is its luminous intensity, as given by Eq. 8.13, then

$$A_{ik} = \frac{L_{ik}}{\hbar \omega} = \frac{4\omega_{ik}^3}{3\hbar c^3} \langle ed \rangle_{ik}^2 \tag{8.15}$$

must be the *probability* that the atom emits a photon during one unit of time. Since the emission of a photon is accompanied by a transition from the initial state of the atoms ψ_k to the final state ψ_i, A_{ik} is usually called the **transition probability**. The lifetime of the initial (excited) state is in this case given by

$$\tau = \frac{1}{A_{ik}} \tag{8.16}$$

---

[3] This so-called *second quantization* would lead to the correct theory—the quantum field theory of light emission.

if the final state ψ_i is the only one to which ψ_k can decay. If there are several final states possible, the lifetime of the state ψ_k is obviously given by[4]

$$\tau = \frac{1}{\sum\limits_{i=1} A_{ik}} \tag{8.17}$$

We mention here in passing that we can also make a strictly classical theory of light emission. In this theory the light emission from a classical oscillator—an elastically bound electron—is calculated, using classical electrodynamics. If a classical oscillator radiates energy, its oscillation must be damped as a result of the energy loss. The *radiation damping* leads to an exponential decay of the luminous intensity, and the decay constant γ can be calculated. We obtain

$$\gamma = \frac{2e^2\omega^2}{mc^3} \tag{8.18}$$

where m is the electron mass. Obviously, in the classical as well as in the quantum-mechanical case the luminous intensity is proportional to the decay constant. The ratio

$$\frac{A_{ik}}{\gamma} = \frac{2\omega_{ik}m\langle d\rangle^2}{3\hbar} = f_{ik} \tag{8.19}$$

is called the **oscillator strength** and gives the luminous intensity of the atom in units of the luminous intensity of a classical oscillator. The value given for the transition probability by Eq. 8.15 agrees in first approximation with the one obtained from quantum electrodynamics. The higher order terms are so small that we cannot detect their presence with current experimental techniques. We return to Eq. 8.15, which we write in the form:

$$A_{ik} = \frac{4\omega_{ik}^3}{3\hbar c^3} |(\psi_i, e\mathbf{d}\psi_k)|^2 \propto |(\psi_i, e\mathbf{d}\psi_k)|^2 \tag{8.20}$$

The coefficients $(\psi_i, e\mathbf{d}\psi_k)$ fill a two-dimensional array and are, therefore, called **matrix elements**. We shall hear more about them later. The structure of Eq. 8.20 is of more than passing interest. Whenever transition probabilities are calculated, the result can be written in the form

$$T_{ik} \propto |(\psi_i, Q\psi_k)|^2 \tag{8.21}$$

where ψ_k and ψ_i are the wave functions of the initial and final state, and Q is the operator of the perturbation (here the electric dipole moment) that causes the transition from one state to the other. Initial and final states can be of a very general nature, and Table 8.1 may serve to illustrate this point. We

[4] See Problem 8.7.

Table 8.1

Process	Initial state	Final state
Light emission	Excited atom	Atom in ground state + photon
Radioactive decay	Nucleus N	Nucleus N' + α-particle, etc.
Scattering	Proton moving toward a nucleus + nucleus at rest	Proton moving away from nucleus + recoiling nucleus

mention here that if the transition from the state ψ_k to the state ψ_i has a large probability, then the probability $T_{ki} \propto |(\psi_k, Q\psi_i)|^2$ of the inverse transition is also large. The reader should be able to prove this statement, assuming that the operator Q is hermitian. (see Problem 8.3)

8.2   THE SELECTION RULE FOR *l*

The actual calculation of atomic transition probabilities can be a very tedious business. We shall not attend to it here and settle instead for the derivation of selection rules. A **selection rule** is a rule that tells one whether or not a certain transition *can occur* without making any statement about the numerical value of the transition probability.

First we consider the quantum number *l*. The wave functions of the hydrogen atom, or for that matter of any atom with a central potential, are of the form:

$$u(\mathbf{r}) = \chi(r)P_l^m(\vartheta)e^{im\varphi} \tag{8.22}$$

A change from $\mathbf{r}$ to $-\mathbf{r}$ corresponds to a change from ϑ to $\pi - \vartheta$, and from φ to $\pi + \varphi$ (see Figure 5.2) in a spherical polar coordinate system. Since

$$\sin(\pi - \vartheta) = \sin\vartheta; \quad \cos(\pi - \vartheta) = -\cos\vartheta; \quad \text{and} \quad e^{i(\varphi+\pi)} = -e^{i\varphi} \tag{8.23}$$

an inspection of the hydrogen wave functions (Table 5.1) shows that all functions with even values of *l* have even parity whereas all those with odd values of *l* have odd parity. Considering that an integral vanishes between symmetrical limits if the integrand has odd parity, and considering that the operator of the electric dipole moment has odd parity, we can state the following **selection rule**:

The expectation value of the electric dipole moment—and with it the transition probability—vanishes unless initial and final state have different parity, i.e., unless

$$l_{\text{initial}} - l_{\text{final}} = \Delta l \neq 0, 2, 4, \ldots \qquad (8.24)$$

This necessary but not sufficient condition for light emission is known as **Laporte's rule.** A straightforward but tedious calculation shows that actually an even more restrictive selection rule applies:

$$\Delta l = \pm 1 \qquad (8.25)$$

8.3   THE SELECTION RULES FOR m_l

Now we consider the quantum number m. Since the states of equal n and l but different m are degenerate we assume that a magnetic field in the z-direction is present. This field is assumed to be so small that it does not appreciably change the energy levels or the transition probability. The field is merely there to assure that the m-degeneracy is lifted; therefore, we have a right to assign to the atom a definite value of the quantum number m. First, we calculate the light emission in the x-y plane which is due to a dipole moment in the z-direction.

$$\langle ez \rangle_{ik} = \int_0^\infty \chi_i \chi_k r^3 \, dr \int_0^\pi P_{l_i}^{m_i} P_{l_k}^{m_k} \sin \vartheta \cos \vartheta \, d\vartheta \int_0^{2\pi} e^{-im_i \varphi} e^{im_k \varphi} \, d\varphi \quad (8.26)$$

We take a closer look at the last integral:

$$\int_0^{2\pi} \exp\left[i(m_k - m_i)\varphi\right] d\varphi = \int_0^{2\pi} e^{i\Delta m\varphi} \, d\varphi \qquad (8.27)$$

This integral vanishes unless $\Delta m = 0$. Hence, in the presence of a magnetic field in the z-direction, linearly polarized light can be observed[5] at a right angle to the field, resulting from transitions with

$$\Delta m = 0 \qquad (8.28)$$

Notice that this rule has been derived neglecting the spin, and is thus only applicable to transitions between spinless (i.e., singlet) states. We state here without proof that it is valid in the presence of a spin in the more general form:

$$\Delta m_J = 0 \qquad (8.29)$$

Now we form

$$x \pm iy = r(\sin \vartheta \cos \varphi \pm i \sin \vartheta \sin \varphi) = r \sin \vartheta e^{\pm i\varphi} \qquad (8.30)$$

---

[5] This is only a necessary condition for the emission of light. It could, of course, be that the ϑ- or the r-dependent integral vanishes.

From this we derive the expectation value:

$$\langle x \pm iy \rangle = \int (\cdots)\, dr \int (\cdots)\, d\vartheta \int_0^{2\pi} e^{i(m_k - m_i \pm 1)\varphi}\, d\varphi \tag{8.31}$$

Hence

$$\langle x \pm iy \rangle = 0 \tag{8.32}$$

or

$$\langle x \rangle = \langle y \rangle = 0 \tag{8.33}$$

unless

$$m_k = m_i \mp 1 \tag{8.34}$$

Hence the dipole moment in the x-y plane can lead to the emission of light in the field direction *if*

$$\Delta m = \pm 1 \tag{8.35}$$

To determine the polarization of the light emitted in the z-direction, we recall that, according to Eq. 8.9, a factor

$$\exp\left[\frac{i(E_i - E_k)t}{\hbar}\right] \tag{8.36}$$

appears in the expression for the electric dipole moment. If we consider that $i = e^{i\pi/2}$ and multiply this into the time-dependent part (Eq. 8.36), we see that according to Eq. 8.31 the x- and the y-component of the dipole moment are out of phase by $\pi/2$. The light emitted in field direction must, therefore, be circularly polarized.

It should be mentioned that nature has ways to get around the above selection rules. Even if the matrix element of the electric dipole moment vanishes, an atom will always go to the ground state (eventually), utilizing processes having operators with parity and angular momentum properties different from those of the electric dipole operator. Possible mechanisms by which such forbidden transitions can occur are the following.

Magnetic Dipole Radiation

If electric dipole radiation is the quantum-mechanical analog of the radiation from a dipole antenna, magnetic dipole radiation is the quantum-mechanical analog of the radiation from a solenoid. The transition probabilities for magnetic dipole transitions are usually very much smaller than those for electric dipole transitions.

Higher Electric Multipole Radiation

Even if two states are not connected by an electric or magnetic dipole moment, they can nevertheless have an oscillating electric quadrupole, etc., moment. Again, the transition probability for such transitions is very much smaller than that for electric dipole radiation.

Two Quantum Transitions

The $2s$ state of hydrogen cannot decay to the $1s$ state because that would imply $\Delta l = 0$. There are no other electric or magnetic moments to help the good cause and, therefore, the atom is left high and dry once it is in the $2s$ state. Usually deexcitation will occur when the atom collides with another atom but, even in a perfect vacuum, the atom can go to the ground state by the simultaneous emission of two photons. Since two-quantum emission is very much less probable than one-quantum emission, the $2s$ state is said to be **meta-stable**.

PROBLEMS

8.1 Prove that

$$\int_{-a}^{a} f(x)\, dx = 0$$

if $f(x)$ has odd parity.

8.2 Calculate the probability of the electric dipole transition between the states $(n, l, m) = (2, 1, 0)$ and $n = 1$ of the hydrogen atom. Neglect any effects that result from the electron or proton spin.

8.3 Quantum-mechanical transitions between two states can take place in both directions (for example, an atom which can emit a photon in going from a state ψ_k to a state ψ_i can also absorb a photon in the state ψ_i and go to the state ψ_k). Show that the transition probabilities for these two processes are always proportional to each other.

8.4 Name all the states of the He atom that have $n = 2$. Which of these states are meta-stable?

8.5 Give more examples of quantum-mechanical transitions in the spirit of Table 8.1. List also the inverse transitions.

8.6 The oscillator strength of the $2p - 1s$ transition of the hydrogen atom is $f = .139$. What is the transition probability? What is the lifetime of the $2p$ state? Assume the elementary charge to be twice as big as it is. How would the lifetime of the $2p$ state change?

8.7 Verify Eq. 8.17.

8.8 A muonic carbon atom is a carbon atom in which a negative muon "orbits" the nucleus far inside the innermost electron shell. Using the value of the transition probability for the $2p - 1s$ transition, calculated in Problem 8.6, find the lifetime of the $2p$ state of the muonic carbon atom.

Solution of 8.3. The probability for a quantum-mechanical transition between two states ψ_i, ψ_k is given by an expression of the form

$$P \propto |(\psi_k, Q\psi_i)|^2$$

where Q is the operator of the interaction that brings about the transition (for instance the electric dipole operator in the case of light emission). The probability for the inverse transition from the state ψ_k to the state ψ_i under the influence of the same operator Q is then

$$P' \propto |(\psi_i, Q\psi_k)|^2$$

The operator Q, describing a physical quantity, must be a hermitian operator. Hence

$$(\psi_i, Q\psi_k) = (Q\psi_i, \psi_k) = (\psi_k, Q\psi_i)^*$$

Hence

$$|(\psi_i, Q\psi_k)|^2 = |(\psi_k, Q\psi_i)|^2 \quad \text{or} \quad P \propto P'$$

We have written $P' \propto P$ rather than $P' = P$ since more than the absolute square of the matrix element enters into P'. For instance, let P be the probability that an atom spontaneously goes from an excited state to the ground state, emitting a photon. The probability of the opposite process, excitation of the atom by resonance absorption of a photon, will in this case depend on the absolute square of the same matrix element, but it will also depend on the abundance of photons available for absorption (i.e., on the intensity of the incident light).

9

MATRIX MECHANICS

In the preceding chapters we have formulated quantum mechanics with the help of partial differential equations. The differential equations were to be solved for wave functions which by themselves did not represent any measurable quantities. Dynamic variables came out of this theory in the form of eigenvalues. The fundamentals of this approach to quantum mechanics are due to Schrödinger. There exists another formulation of the theory that is due to Heisenberg. The Heisenberg approach spurns quantities that cannot be measured directly, such as probability amplitudes. Instead, measurable quantities are directly related with each other through the rules of matrix algebra. The two approaches yield the same results and are—no matter how different they look—very closely related mathematically. The rules of matrix mechanics were derived directly from the experimental results by Heisenberg. Only later was it recognized by Schrödinger how closely akin mathematically his and Heisenberg's forms of the theory really were. Having already mastered Schrödinger's *wave mechanics*, we shall use it as a convenient access road to Heisenberg's *matrix mechanics*.

Although both theories are completely equivalent it turns out that in some applications one is more convenient to use than the other. In the following chapters we shall concentrate on those problems that lend themselves more readily to a solution with matrix methods.

9.1 MATRIX ELEMENTS

Previously we had defined the expectation value of a linear operator Q as

$$\langle Q \rangle_i = \int \psi_i^* Q \psi_i \, d\tau = (\psi_i, Q \psi_i) \tag{9.1}$$

This expectation value was defined as the average of many measurements of the quantity described by the operator Q. The measurements were to be made while the system was in a state described by a *set* of quantum numbers symbolized here by the one subscript i. Now we expand ψ_i^* and ψ_i in terms of eigenfunctions of another linear operator (for instance, in terms of the eigenfunctions u_k of the Hamiltonian of the system).

$$\psi_i^* = \sum_{k=1} c_{ik}^* u_k^* \qquad \text{and} \qquad \psi_i = \sum_{m=1} c_{im} u_m \qquad (9.2)$$

Hence

$$(\psi_i, Q\psi_i) = \sum_{k=1} \sum_{m=1} c_{ik}^* c_{im} \int u_k^* Q u_m \, d\tau \qquad (9.3)$$

The definite integrals $\int u_k^* Q u_m \, d\tau$ form a two-dimensional array of numbers. We encountered such an array in the previous chapter and for no good reason, called the numbers **matrix elements**:

$$q_{km} = (u_k, Q u_m) \qquad (9.4)$$

Now we shall show that they deserve this name because the rules of matrix algebra apply to them. First we show that the rule of **matrix addition**

$$(A + B)_{ik} = a_{ik} + b_{ik} \qquad (9.5)$$

applies. Let Q and P be two linear operators. Then

$$(Q + P)_{ik} = \int u_i^*(Q + P)u_k \, d\tau = \int u_i^* Q u_k \, d\tau + \int u_i^* P u_k \, d\tau = q_{ik} + p_{ik} \qquad (9.6)$$

Matrix multiplication is defined by

$$(AB)_{ik} = \sum_{m=1}^{n} a_{im} b_{mk} \qquad (9.7)$$

To show that the same law applies to our two-dimensional arrays, we form

$$\varphi_k = Q u_k \qquad (9.8)$$

where Q is hermitian, and expand it in terms of the u_i:

$$Q u_k = \varphi_k = \sum_{i=1} c_{ki} u_i \qquad (9.9)$$

We multiply from the left with u_l^* and integrate

$$(u_l, Q u_k) = q_{lk} = \sum_{i=1} c_{ki}(u_l, u_i) \qquad (9.10)$$

Since the eigenfunctions of a hermitian operator are orthonormal,[1] it follows that

$$q_{lk} = c_{kl} \qquad (9.11)$$

Hence

$$Q u_k = \sum_{l=1} q_{lk} u_l \qquad (9.12)$$

[1] Or can be made orthonormal.

Therefore

$$(u_i, PQu_k) = \left(u_i, P\sum_{l=1} q_{lk}u_l\right) = \sum_{l=1} p_{il}q_{lk} \qquad \text{q.e.d.} \qquad (9.13)$$

The associative and distributive law are obviously satisfied since the linear operators satisfy them. From the above definitions, the theorem follows.

THEOREM

If the u_k form a complete set of functions and Q is a hermitian operator, then the matrix $\mathbf{Q}$ is hermitian.

Proof

$$q_{ik} = (u_i, Qu_k) = (Qu_i, u_k) = (u_k, Qu_i)^* = q_{ki}^* \qquad (9.14)$$

Finally, a simple but important theorem.

THEOREM

The matrix elements of an operator taken between the eigenfunctions of this operator form a diagonal matrix.

Proof (trivial)

As an example of this last theorem we write down the elements of the matrix of the Hamiltonian:

$$(u_i, Hu_k) = (u_i, E_k u_k) = E_k(u_i, u_k) = E_k \qquad (9.15)$$

We sum up: For every hermitian operator we can form a hermitian matrix, using an arbitrary complete set of functions. If these functions happen to be the eigenfunctions of the hermitian operator, the resulting matrix is diagonal and the matrix elements are the eigenvalues of the operator. Operators and matrices are related in much the same way as are vectors and their representatives. A *vector* exists in its own right without reference to any *coordinate system*. We can *represent* the vector by a *column*

$$\mathbf{r} = \begin{pmatrix} x \\ y \\ z \end{pmatrix}$$

giving its projections in a *particular coordinate system*. Similarly, an operator is a mathematical entity that exists without reference to any system of functions one might care to apply it to. We can *represent* the operator Q in a *particular set of functions u_k* by forming all the matrix elements.

$$q_{ik} = (u_i, Qu_k) \qquad (9.16)$$

A quantum-mechanical problem is solved if we have found the eigenvalues and eigenfunctions of the Hamiltonian H and of those other operators Q in which we might have been interested. We can then construct the diagonal matrices, whose elements are the expectation values, and the nondiagonal matrices, whose elements $q_{ik} = (\psi_i, Q\psi_k)$ are related to the transition probabilities between the eigenstates ψ_i and ψ_k of the Hamiltonian. Now the question is: Do we have to solve the Schrödinger equation in order to form this diagonal matrix? The answer is no. We shall show in the next paragraph that there exists a strong algebraic connection between the different matrix representations of the same operator. This will enable us to form a matrix representation of an operator in *any* complete set of functions and then to find its one and only diagonal matrix representation by means of a unitary transformation.

9.2 THE SOLUTION OF A QUANTUM MECHANICAL PROBLEM BY MEANS OF A UNITARY TRANSFORMATION

We start with two complete sets of functions:

$$\varphi_k, \psi_k \qquad k = 1, 2, 3, \ldots$$

Obviously, we can expand one in terms of the other:

$$\varphi_k = \sum_{i=1} a_{ki}\psi_i \tag{9.17}$$

The expansion coefficients a_{ki} form a matrix. Similarly,

$$\psi_i = \sum_{n=1} b_{in}\varphi_n \tag{9.18}$$

Substitution of Eq. 9.18 into Eq. 9.17 yields

$$\varphi_k = \sum_{i=1} \sum_{n=1} a_{ki}b_{in}\varphi_n \tag{9.19}$$

The coefficients of φ_n:

$$\sum_{i=1} a_{ki}b_{in}$$

must satisfy

$$\sum_{i=1} a_{ki}b_{in} = \delta_{kn}\begin{cases} = 1 & \text{if} \quad k = n \\ = 0 & \text{if} \quad k \neq n \end{cases} \tag{9.20}$$

Hence

$$\mathbf{AB} = \mathbf{1} \qquad \text{or} \qquad \mathbf{B} = \mathbf{A}^{-1} \tag{9.21}$$

We shall now show that the matrices $\mathbf{A}$ and $\mathbf{B}$ are unitary.

Proof

$$(\varphi_i, \varphi_k) = \delta_{ik} = \left(\sum_{l=1} a_{il}\psi_l, \sum_{n=1} a_{kn}\psi_n \right)$$

$$= \sum_{l=1}\sum_{n=1}(a_{il}\psi_l, a_{kn}\psi_n) = \sum_{l=1}\sum_{n=1} a_{il}{}^*a_{kn}\delta_{ln} = \sum_{l=1} a_{il}{}^*a_{kl} = \sum_{l=1} a_{kl}a_{li}{}^\dagger$$

$$(9.22)$$

Written in matrix form, Eq. 9.22 reads

$$\mathbf{A}\mathbf{A}\dagger = \mathbf{1} \qquad (9.23)$$

We have thus found the theorem:

THEOREM

Any complete set of functions can be transformed into another complete set by means of a unitary transformation.

Next we see what the transformation **A**, defined as above, will do to the matrix representation of an operator Q. Let **Q** be a matrix whose elements are

$$q_{ik} = (\psi_i, Q\psi_k) \qquad (9.24)$$

Let Q' be the matrix representation of the same operator Q in another complete set of functions:

$$q_{ik}{}' = (\varphi_i, Q\varphi_k) \qquad (9.25)$$

Using a unitary transformation, we express the $q_{ik}{}'$ with the help of the ψ_k:

$$q_{ik} = \sum_{n=1}\sum_{l=1}(a_{in}\psi_n, Qa_{kl}\psi_l)$$

$$= \sum_{n=1}\sum_{l=1} a_{in}{}^*q_{nl}a_{kl} = \sum_{n=1}\sum_{l=1} a_{in}{}^*q_{nl}a_{lk}\dagger^* \qquad (9.26)$$

or in matrix notation

$$\mathbf{Q}' = \mathbf{A}^*\mathbf{Q}\mathbf{A}\dagger^* \qquad (9.27)$$

We have thus found the following theorem.

THEOREM

The same unitary transformation that transforms one complete set of functions into another complete set through

$$\varphi = \mathbf{A}\psi \qquad \text{*or written more explicitly*} \qquad \varphi_k = \sum_i a_{ki}\psi_i \qquad k = 1, 2, 3, \dots$$

$$(9.28)$$

transforms the matrix representation of an operator in one set into the representation in the other set through

$$\mathbf{Q}' = \mathbf{A}^*\mathbf{Q}\mathbf{A}\dagger^* \qquad (9.29)$$

Therefore, it should be possible to find the diagonal matrix representation of an operator (i.e., the representation that has the eigenvalues of the operator in the main diagonal) in the following way:

(a) Write down the matrix representation of the operator in some arbitrary complete set of functions φ_k.

(b) Find the unitary transformation that transforms this matrix into a diagonal matrix.

(c) Use the same transformation to transform the arbitrary set of functions φ_k into the complete set of eigenfunctions ψ_k of the operator.

If this procedure is unique, it should solve the problem completely since it gives us the eigenfunctions and the eigenvalues of the operator, and from these two quantities we can calculate all there is to be known. Now, we shall show that there is one and only one, diagonal matrix representation of an operator by giving a unique procedure for finding it. Let H' be a matrix representation of the Hamilton operator in some arbitrary complete set of functions φ_k:

$$H_{ik}' = (\varphi_i, H\varphi_k) \tag{9.30}$$

We want to find the diagonal representation $\mathbf{H} = \mathbf{E}$ with the elements[2]

$$(\psi_i, H\psi_k) = E_{ik}\delta_{ik} \tag{9.31}$$

where the ψ_k are the eigenfunctions and the E_k the energy eigenvalues of the Hamilton operator. The unitary transformation that transforms $\mathbf{H}'$ into the diagonal matrix $\mathbf{E}$ is defined by

$$A\dagger^*H'A^* = E \tag{9.32}$$

Multiplication from the left with A^* yields

$$H'A^* = A^*E \tag{9.33}$$

or

$$\sum_{m=1} H_{im}' a_{mk}^* = \sum_{m=1} a_{im}^* E_{mk} = a_{ik}^* E_k \tag{9.34}$$

For any value of k, Eq. 9.34 is a set of coupled linear homogeneous equations for the a_{ik}:

$$i = 1: H_{11}' a_{1k}^* + H_{12}' a_{2k}^* + H_{13}' a_{3k}^* + \cdots = a_{1k}^* E_k$$
$$i = 2: H_{21}' a_{1k}^* + H_{22}' a_{2k}^* + H_{23}' a_{3k}^* + \cdots = a_{2k}^* E_k \tag{9.35}$$
$$i = 3: \cdots, \text{etc.}, \ldots$$

[2] We conform, in general, with the convention that we have used for matrices: (1) Capital letters for operators and (2) lower case letters for matrix elements. We depart from this convention only when—as in the case of energy eigenvalues—the use of capital letters for particular eigenvalues is well established.

These linear homogeneous equations for the a_{ik}^* can be solved if, and only if, the secular determinant vanishes:

$$\begin{vmatrix} (H_{11}' - E_k) & H_{12}' & H_{13}' & \cdots \\ H_{21}' & (H_{22}' - E_k) & H_{23}' & \cdots \\ H_{31}' & H_{32}' & (H_{33}' - E_k) & \cdots \\ \cdot & \cdot & \cdot & \\ \cdot & \cdot & \cdot & \\ \cdot & \cdot & \cdot & \end{vmatrix} = 0 \qquad (9.36)$$

This determinant is an equation of order n, $(n \to \infty)$. It has n possible solutions, the E_k, that need not all be different (degeneracy). We have thus shown that there exists, at least in principle, a unique way to determine the eigenvalues E_k and all the coefficients of the unitary transformation **A** that transforms the φ_k into the ψ_k. This method of solving a quantum-mechanical problem is completely equivalent to solving the Schrödinger equation. Since it involves the solution of an infinite set of coupled linear equations when applied to a problem with infinitely many different eigenfunctions, it is not often used for problems of this type. The real advantage of the method becomes apparent if we apply it to perturbation problems, as we shall do in the next chapter.

9.3 THE ANGULAR MOMENTUM MATRICES

For later use we shall now derive matrix representations of the angular momentum operators. It will turn out that we can do this without ever specifying a particular complete set of functions. We start from the algebraic (commutation) relations between the operators that we derived in Chapter 6.4.

Since the algebraic relations between operators remain valid for their matrix representations, we know that the commutators[3]

$$(a)[\mathbf{L}_x, \mathbf{L}_y] = i\mathbf{L}_z, \qquad (b)[\mathbf{L}^2, \mathbf{L}_x] = 0 \qquad (9.37)$$

$$(a)[\mathbf{L}_y, \mathbf{L}_z] = i\mathbf{L}_x, \qquad (b)[\mathbf{L}^2, \mathbf{L}_y] = 0 \qquad (9.38)$$

$$(a)[\mathbf{L}_z, \mathbf{L}_x] = i\mathbf{L}_y, \qquad (b)[\mathbf{L}^2, \mathbf{L}_z] = 0 \qquad (9.39)$$

must also be valid for the angular momentum matrices. Since two diagonal matrices always commute it follows from Eqs. 9.37, 9.38a, and 9.39a that no two of the component matrices $\mathbf{L}_x$, $\mathbf{L}_y$, $\mathbf{L}_z$ can be simultaneously diagonal.

---

[3] For convenience, we follow the convention of expressing angular momenta in units of $\hbar$, this amounts to setting $\hbar = 1$ in our equations.

We know that the angular momentum matrices are hermitian because the corresponding operators are hermitian. We choose our representation so that L_z and L^2 are diagonal and set out to find L_x, L_y, L_z and L^2 in this particular representation. It will turn out that the assumption that L_z and L^2 are diagonal, together with the commutation relations Eqs. 9.37 to 39, suffices to determine the elements of the angular momentum matrices unambiguously.

For later use we introduce the *nonhermitian* matrix

$$\mathbf{L}_+ = \mathbf{L}_x + i\mathbf{L}_y \qquad (9.40)$$

and its hermitian conjugate[4]

$$\mathbf{L}_+{}^\dagger = \mathbf{L}_- = \mathbf{L}_x - i\mathbf{L}_y \qquad (9.41)$$

These matrices satisfy the following commutation relations:

$$[\mathbf{L}_z, \mathbf{L}_+] = [\mathbf{L}_z, \mathbf{L}_x] + i[\mathbf{L}_z, \mathbf{L}_y] = i\mathbf{L}_y + \mathbf{L}_x = \mathbf{L}_+ \qquad (9.42)$$

Also

$$[\mathbf{L}_-, \mathbf{L}_z] = [\mathbf{L}_x, \mathbf{L}_z] - i[\mathbf{L}_y, \mathbf{L}_z] = \mathbf{L}_- \qquad (9.43)$$

and

$$[\mathbf{L}_+, \mathbf{L}_-] = -i[\mathbf{L}_x, \mathbf{L}_y] + i[\mathbf{L}_y, \mathbf{L}_x] = 2\mathbf{L}_z \qquad (9.44)$$

and finally (and obviously)

$$[\mathbf{L}^2, \mathbf{L}_+] = [\mathbf{L}^2, \mathbf{L}_-] = 0 \qquad (9.45)$$

since both L_x and L_y commute with L^2. We can also easily verify that

$$\mathbf{L}_+\mathbf{L}_- = \mathbf{L}^2 - \mathbf{L}_z{}^2 + \mathbf{L}_z \qquad (9.46)$$

and

$$\mathbf{L}_-\mathbf{L}_+ = \mathbf{L}^2 - \mathbf{L}_z{}^2 - \mathbf{L}_z \qquad (9.47)$$

We still do not know what the physical significance of L_+ and L_- is. To find out, we investigate the *operators* L_+ and L_-. Let $\psi_{\lambda m}$ be a simultaneous eigenfunction of the operators L^2 and L_z with the eigenvalues λ and m:

$$L^2\psi_{\lambda m} = \lambda\psi_{\lambda m}; \qquad L_z\psi_{\lambda m} = m\psi_{\lambda m} \qquad (9.48)$$

We know that such simultaneous eigenfunctions must exist since the operators L^2 and L_z commute. In the case of the hydrogen eigenfunctions we had found that $\lambda = l(l + 1)$. We have as yet no right to assume that λ will be of the same form for any eigenfunction $\psi_{\lambda m}$ of the operator L^2 and, hence, use the noncommittal form λ for the eigenvalues of L^2. We apply L_zL_+ to $\psi_{\lambda m}$. According to Eq. 9.42 this yields

$$L_z(L_+\psi_{\lambda m}) = L_+L_z\psi_{\lambda m} + L_+\psi_{\lambda m}$$
$$= L_+m\psi_{\lambda m} + L_+\psi_{\lambda m} = L_+(m + 1)\psi_{\lambda m} = (m + 1)(L_+\psi_{\lambda m}) \qquad (9.49)$$

---

[4] Of course, the hermitian conjugate of a non-hermitian matrix is also a non-hermitian matrix. The "non-hermitian hermitian conjugate" is only a linguistic curiousity, not a mathematical one.

Comparing the left and the right side of Eq. 9.49, we see that $L_+\psi_{\lambda m}$ is an eigenfunction of L_z with the eigenvalue $m + 1$. Since $\psi_{\lambda m}$ is an eigenfunction of L_z with the eigenvalue m this means that L_+ when applied to one of the functions $\psi_{\lambda m}$ raises the value of its quantum number m by one. Therefore, it is often called a *raising operator*. Similarly we can show that L_- is a *lowering operator*, i.e., it lowers the value of m by one. It should be pointed out that L_+ raises and L_- lowers the quantum number m regardless of what else it does to the eigenfunction of L^2 and L_z. Unless $m + 1$ is a nondegenerate eigenvalue of L_z, we are not justified in assuming that

$$L_+\psi_{\lambda m} = \psi_{\lambda m+1} \tag{9.50}$$

but only that

$$L_+\psi_{\lambda m} = \psi'_{\lambda m+1} \tag{9.51}$$

where ψ' is a function that may, or may not, differ from ψ. From Eq. 9.45 follows

$$L^2 L_+\psi_{\lambda m} = L_+ L^2\psi_{\lambda m} = L_+ \lambda\psi_{\lambda m} = \lambda L_+\psi_{\lambda m} \tag{9.52}$$

In other words, while L_+ and L_- raise or lower the quantum number m by one, they leave the quantum number λ untouched. Next we apply L_+ several times in succession to an eigenfunction $\psi_{\lambda m}$

$$L_+ L_+ \cdots L_+\psi_{\lambda m} \tag{9.53}$$

thus, raising m more and more. Now, m is the expectation value of the z-component of the angular momentum vector and λ the expectation value of the square of its absolute value. Therefore, the above procedure must lead us to some highest value of m which we shall call j since the z-component of a vector cannot be larger than the entire vector. Hence, there must be an eigenfunction $\psi_{\lambda j} \neq 0$ and eigenfunctions with $m \geqslant j + 1$ must not exist. Hence

$$L_+\psi_{\lambda j} = 0 \tag{9.54}$$

Similarly, a consecutive application of L_- must lead to some lowest value of $m = j'$ so that

$$L_-\psi_{\lambda j'} = 0 \qquad \text{for} \qquad \psi_{\lambda j'} \neq 0 \tag{9.55}$$

The difference $j - j'$ must, of course, be a positive integer. Now we apply $L_+ L_-$ and $L_- L_+$ to $\psi_{\lambda j'}$ and $\psi_{\lambda j}$, respectively,

$$L_+ L_-\psi_{\lambda j'} = 0 \qquad \text{and} \qquad L_- L_+\psi_{\lambda j} = 0 \tag{9.56}$$

Using Eqs. 9.46 and 9.47 we obtain

$$L_+ L_-\psi_{\lambda j'} = (\lambda - j'^2 + j')\psi_{\lambda j'} = 0 \qquad \text{and} \qquad L_- L_+\psi_{\lambda j}$$
$$= (\lambda - j^2 + j)\psi_{\lambda j} = 0 \tag{9.57}$$

Since

$$\psi_{\lambda j'} \neq 0 \neq \psi_{\lambda j}$$

it follows from Eq. 9.57 that

$$j'(j' - 1) = \lambda \qquad \text{and} \qquad j(j + 1) = \lambda \qquad (9.58)$$

Equation 9.58, together with the fact that

$$j - j' = k \qquad (9.59)$$

is a positive integer, allows us to determine k. The quadratic equation

$$(j - k)(j - k - 1) = j^2 + j \qquad (9.60)$$

has two solutions:

$$k = -1 \qquad \text{and} \qquad k = 2j \qquad (9.61)$$

The first one has to be discarded because it is negative. From the second it follows that

$$j' = -j \qquad (9.62)$$

or

$$\lambda = j(j + 1) \qquad (9.63)$$

and

$$m = -j, -(j - 1), -(j - 2), \ldots, j - 2, j - 1, j \qquad (9.64)$$

In other words, what we found for the special case of the hydrogen eigenfunctions applies to any wave function $\psi_{\lambda m}$ that is an eigenfunction to L^2 and L_z:

(a) the eigenvalue of L^2 can be expressed as $j(j + 1)$, and
(b) m_j can assume any of $2j + 1$ values from $-j$ to $+j$.

Now comes the *big difference*: According to Eq. 9.61, $k = 2j$ is an integer, which means that j can have integer or half-integer values.[5] In other words,

$$j = 0, \tfrac{1}{2}, 1, \tfrac{3}{2}, 2, \ldots, \text{etc.} \qquad (9.65)$$

We have, thus, justified all the assumptions made earlier in the discussion of the vector model. In the beginning of this section we promised that the elements of the angular momentum matrices could be determined from the commutation relations and the assumption that L^2 and L_z are diagonal. We have yet to make good on this promise.

9.4 AN EXPLICIT REPRESENTATION OF THE ANGULAR MOMENTUM MATRICES

To find the elements of the angular momentum matrices we apply L_+ to some eigenfunction[6] ψ_{jm}:

$$L_+\psi_{jm} = \psi'_{jm+1} \qquad (9.66)$$

---

[5] This does not prove the existence of the spin but shows that spin, if it exists, can be described within the framework of our formalism.
[6] Instead of $\lambda = j(j + 1)$ we shall use j from now on to label the eigenfunctions.

We know that ψ'_{jm+1} is an eigenfunction to L_z with the eigenvalue $m+1$. If $m+1$ is a nondegenerate eigenvalue, we must have

$$\psi'_{jm+1} = a_m \psi_{jm+1} \tag{9.67}$$

where a_m is a constant. Since we have shown in the preceding paragraph that there are as many different eigenvalues m as there are eigenfunctions ψ_{jm} (namely, $2j+1$), we conclude that the eigenvalues m are nondegenerate. Therefore

$$L_+ \psi_{jm} = a_m \psi_{jm+1} \tag{9.68}$$

Similarly we find that

$$L_- \psi_{jm} = b_m \psi_{jm-1} \tag{9.69}$$

Since the ψ_{jm} form an orthonormal set, the matrix L_+ can only have nonvanishing elements of the form:

$$(\psi_{jm+1} L_+ \psi_{jm}) = a_m \tag{9.70}$$

(i.e., only the elements whose row index is one higher than their column index differ from zero). The matrix L_+ must, therefore, look like Eq. 9.71.

	$m =$	$\frac{1}{2}$	$-\frac{1}{2}$	1	0	-1	$\frac{3}{2}$	$\frac{1}{2}$	$-\frac{1}{2}$	$-\frac{3}{2}$
$j = \frac{1}{2}$	$\frac{1}{2}$	0	$a_{-\frac{1}{2}}$	0	0	0	0	0	0	0
	$-\frac{1}{2}$	0	0	0	0	0	0	0	0	0
	1	0	0	0	a_0	0	0	0	0	0
$j = 1$	0	0	0	0	0	a_{-1}	0	0	0	0
	-1	0	0	0	0	0	0	0	0	0
	$\frac{3}{2}$	0	0	0	0	0	0	$a_{\frac{1}{2}}$	0	0
$j = \frac{3}{2}$	$\frac{1}{2}$	0	0	0	0	0	0	0	$a_{-\frac{1}{2}}$	0
	$-\frac{1}{2}$	0	0	0	0	0	0	0	0	$a_{-\frac{3}{2}}$
	$-\frac{3}{2}$	0	0	0	0	0	0	0	0	0

$$\tag{9.71}$$

(Note that in accordance with convention we have labeled rows and columns so that m runs from $+j$ to $-j$. This puts the nonvanishing matrix elements *above* the main diagonal.) From Eqs. 9.46 and 9.47 it follows that $\mathbf{L}_+\mathbf{L}_-$ and $\mathbf{L}_-\mathbf{L}_+$ are diagonal matrices. Hence using Eq. 9.41

$$(L_-L_+)_{im} = j(j+1) - m^2 - m = \sum_{k=1} L_{-ik}L_{+km}\delta_{im}$$

$$= \sum L_{-mk}L_{+km} = \sum_{1=k} L^*_{+km}L_{+km} \tag{9.72}$$

Since the only nonvanishing elements of $\mathbf{L}_+$ are those with $k = m + 1$, there is nothing to sum over in Eq. 9.72, and we get

$$L^*_{+m+1,m}L_{+m+1,m} = j(j+1) - m(m+1) \tag{9.73}$$

or (except for a phase factor which we set arbitrarily equal to plus one[7])

$$L_{+m+1,m} = \sqrt{j(j+1) - m(m+1)} \tag{9.74}$$

Letting $j = \frac{1}{2}$, we find that

$$\mathbf{L}_+ = \begin{array}{c|c|c} m & \frac{1}{2} & -\frac{1}{2} \\ \hline \frac{1}{2} & 0 & 1 \\ \hline -\frac{1}{2} & 0 & 0 \end{array} \quad \text{and} \quad \mathbf{L}_- = \begin{array}{c|c|c} m & \frac{1}{2} & -\frac{1}{2} \\ \hline \frac{1}{2} & 0 & 0 \\ \hline -\frac{1}{2} & 1 & 0 \end{array} \tag{9.75}$$

And from

$$\mathbf{L}_x = \tfrac{1}{2}(\mathbf{L}_+ + \mathbf{L}_-) \tag{9.76}$$

it follows that

$$\mathbf{L}_x = \frac{1}{2}\begin{pmatrix} 0 & 1 \\ 1 & 0 \end{pmatrix} \quad \text{and} \quad \mathbf{L}_y = \frac{1}{2}\begin{pmatrix} 0 & -i \\ i & 0 \end{pmatrix} \tag{9.77}$$

Also from the commutator Eq. (9.37a) it follows that

$$\mathbf{L}_z = \frac{1}{2}\begin{pmatrix} 1 & 0 \\ 0 & -1 \end{pmatrix} \tag{9.78}$$

The matrices

$$\sigma_x = \begin{pmatrix} 0 & 1 \\ 1 & 0 \end{pmatrix}, \quad \sigma_y = \begin{pmatrix} 0 & -i \\ i & 0 \end{pmatrix}, \quad \sigma_z = \begin{pmatrix} 1 & 0 \\ 0 & -1 \end{pmatrix} \tag{9.79}$$

are the famous Pauli spin matrices or **Pauli matrices**. In a similar fashion we can find the matrix $\mathbf{L}_+$ for $j = 1$:

$$\mathbf{L}_+ = \sqrt{2}\begin{pmatrix} 0 & 1 & 0 \\ 0 & 0 & 1 \\ 0 & 0 & 0 \end{pmatrix} \tag{9.80}$$

[7] The phase of the angular momentum has no physical significance.

Hence

$$\mathbf{L}_x = \frac{1}{\sqrt{2}} \begin{pmatrix} 0 & 1 & 0 \\ 1 & 0 & 1 \\ 0 & 1 & 0 \end{pmatrix}, \qquad \mathbf{L}_y = \frac{1}{\sqrt{2}} \begin{pmatrix} 0 & -i & 0 \\ i & 0 & -i \\ 0 & i & 0 \end{pmatrix},$$

$$\mathbf{L}_z = \begin{pmatrix} 1 & 0 & 0 \\ 0 & 0 & 0 \\ 0 & 0 & -1 \end{pmatrix} \qquad (9.81)$$

$$\mathbf{L}^2 = 2 \begin{pmatrix} 1 & 0 & 0 \\ 0 & 1 & 0 \\ 0 & 0 & 1 \end{pmatrix} \qquad (9.82)$$

PROBLEMS

9.1 Which of the following statements are correct?
(a) All diagonal matrices are hermitian.
(b) A matrix that is both hermitian and unitary is a diagonal matrix.
(c) All real matrices are hermitian.
(d) A diagonal matrix commutes with any matrix of the same rank.

9.2 Find the elements

$$p_{00} = (u_0, p_x u_0), \qquad p_{01} = (u_0, p_x u_1), \qquad p_{10} = (u_1, p_x u_0)$$

and

$$p_{11} = (u_1, p_x u_1)$$

where u_0 and u_1 are the first two eigenfunctions of the one-dimensional harmonic oscillator and p_x is the x-component of the momentum operator.

9.3 Find the elements

$$\langle qx \rangle_{00} = (u_0, qx u_0), \qquad \langle qx \rangle_{01} = (u_0, qx u_1),$$
$$\langle qx \rangle_{10} = (u_1, qx u_0), \qquad \text{and} \qquad \langle qx \rangle_{11} = (u_1, qx u_1)$$

of the electric dipole moment matrix of the one-dimensional harmonic oscillator.

9.4 Find the matrix element

$$(u_{nlm}, er u_{n'l'm'}) = (u_{100}, er u_{210})$$

for the hydrogen atom.

9.5 Write down the explicit expressions for the raising and lowering operators L_+ and L_- and show, by applying them to the $2p$ eigenfunctions of hydrogen, that they, indeed, raise and lower the value of the quantum number m. (*Hint.* See solution of 9.6.)

9.6 Show explicitly that $L_+ Y_{11} = L_- Y_{1-1}$.

9.7 Derive explicit representations for the matrices $\mathbf{L}_x$, $\mathbf{L}_y$, $\mathbf{L}_z$, and $\mathbf{L}^2$ for $j = \frac{3}{2}$ so that $\mathbf{L}_z$ and $\mathbf{L}^2$ are diagonal.

9.8 Are all the Pauli matrices hermitian? Are they all unitary?

9.9 Demonstrate that the commutation relations derived for the angular momentum operators do, indeed, apply to the angular momentum matrices for $j = \frac{1}{2}$ and $j = 1$.

9.10 Which of the angular momentum matrices for $j = 1$ are hermitian? Which are unitary?

SOLUTIONS

9.2
$$\left(B_0 e^{-x^2/2}, (-i\hbar) \frac{\partial B_0}{\partial x} e^{-x^2/2} \right) = -B_0^2 i\hbar \int_{-\infty}^{\infty} (-x) e^{-x^2} \, dx = 0$$

(Since the integral is taken between symmetrical limits and the integrand has odd parity. See also Problem 8.1)

$$\left(B_0 e^{-x^2/2}, (-i\hbar) \frac{\partial}{\partial x} B_1 2x e^{-x^2/2} \right) = -2i\hbar B_0 B_1 \int_{-\infty}^{\infty} e^{-x^2} (1 - x^2) \, dx$$

$$= -2i\hbar B_0 B_1 \left(\sqrt{\pi} - \frac{\sqrt{\pi}}{2} \right) = -i\hbar B_0 B_1 \sqrt{\pi} = \frac{-i\hbar}{\sqrt{2}}$$

Following this procedure the reader should not find it difficult to evaluate the two remaining matrix elements.

9.6 The raising operator $L_+ = L_x + iL_y$ can be written explicitly with the help of Eqs. 5.104 and 5.105:

$$L_+ = i\hbar \left(\sin \varphi \frac{\partial}{\partial \vartheta} + \cot \vartheta \cos \varphi \frac{\partial}{\partial \varphi} - i \cos \varphi \frac{\partial}{\partial \vartheta} + i \cot \vartheta \sin \varphi \frac{\partial}{\partial \varphi} \right)$$

$$= (i\hbar \sin \varphi + \hbar \cos \varphi) \frac{\partial}{\partial \vartheta} + (i\hbar \cot \vartheta \cos \varphi - \hbar \cot \vartheta \sin \varphi) \frac{\partial}{\partial \varphi}$$

We apply this to

$$Y_{11} \propto \sin \vartheta e^{i\varphi}$$

$$L_+ Y_{11} \propto (i\hbar \sin \varphi + \hbar \cos \varphi) \cos \vartheta e^{i\varphi}$$
$$+ (i\hbar \cot \vartheta \cos \varphi - \hbar \cot \vartheta \sin \varphi) i \sin \vartheta e^{i\varphi}$$

$$= e^{i\varphi} \left\{ \hbar \cos \vartheta \cos \varphi - \hbar \frac{\cos \vartheta}{\sin \vartheta} \cos \varphi \sin \vartheta \right.$$

$$+ i \left(\hbar \sin \varphi \cos \vartheta - \hbar \frac{\cos \vartheta}{\sin \vartheta} \sin \varphi \sin \vartheta \right) \right\} = 0$$

Analogously, we may show that

$$L_- Y_{1-1} = 0$$

10

STATIONARY PERTURBATION THEORY

10.1 THE PERTURBATION OF NONDEGENERATE STATES

We have already mentioned that most quantum-mechanical problems must be solved with one or the other of a variety of approximative methods. One of these methods is due to Schrödinger and deals with the following situation:

The state of a system is mainly determined by some strong interaction[1] (for instance, the Coulomb interaction between the electrons and the nucleus). The Hamiltonian $H^{(0)}$ related to this interaction is known and so are its eigenfunctions[2] and eigenvalues:

$$H^{(0)}\psi_k^{(0)} = E_k^{(0)}\psi_k^{(0)} \tag{10.1}$$

In addition, there exist weaker[1] interactions which result in *small changes* of the Hamiltonian, its eigenfunctions, and eigenvalues. As examples we mention the interaction with an external electric or magnetic field, and the interaction between spin and orbital magnetic moment. Because of the presence of these additional interactions, Eq. 10.1 becomes

$$H\psi_k = E_k\psi_k \tag{10.2}$$

where H, E_k and ψ_k differ slightly from $H^{(0)}$, $E_k^{(0)}$, and $\psi_k^{(0)}$. It is obvious what "differ slightly" means in the case of E_k; however, for H and ψ_k we have to define it more carefully. We assume ψ_k to be of the form

$$\psi_k = \psi_k^{(0)} + \psi_k^{(1)} + \psi_k^{(2)} + \cdots \tag{10.3}$$

and require that $\psi_k^{(1)}$, $\psi_k^{(2)}$, etc., are small compared to $\psi_k^{(0)}$ in some yet to be defined way. As far as H is concerned we assume that it can be written as

$$H = H^{(0)} + H^{(1)} + H^{(2)} + \cdots \tag{10.4}$$

---

[1] The terms strong and weak interaction are usually reserved for the description of certain nonelectromagnetic interactions. We use them here in a broader sense.

[2] We shall see later that we can obtain useful results even without such detailed knowledge of the unperturbed system.

and that $H^{(1)}$, $H^{(2)}$, etc., are small compared to $H^{(0)}$. Again, we have to hedge concerning the exact definition of smallness. The conditions that have to be satisfied by $\psi_k^{(1)}$ and $H^{(1)}$ will evolve in the course of our investigation and it will turn out that the two conditions are closely related. We substitute Eqs. 10.3 and 10.4 into Eq. 10.2 and obtain

$$
\begin{aligned}
H\psi_k &= (H^{(0)} + H^{(1)} + H^{(2)} + \cdots)(\psi_k^{(0)} + \psi_k^{(1)} + \psi_k^{(2)} + \cdots) \\
&= (E_k^{(0)} + E_k^{(1)} + E_k^{(2)} + \cdots)(\psi_k^{(0)} + \psi_k^{(1)} + \psi_k^{(2)} + \cdots) = E_k \psi_k
\end{aligned}
$$

$$(10.5)$$

If we write this explicitly, we obtain an expression containing terms like $H^{(0)}\psi_k^{(0)}$, $H^{(0)}\psi_k^{(1)}$, $H^{(0)}\psi_k^{(2)}$, $H^{(1)}\psi_k^{(0)}$, $H^{(1)}\psi_k^{(1)}$, etc. The sum of the upper indices in these expressions is a measure of their smallness. In other words $H^{(1)}\psi_k^{(1)}$ will be smaller than $H^{(1)}\psi_k^{(0)}$ or $H^{(0)}\psi_k^{(1)}$ if we have defined the smallness of $H^{(1)}$, $H^{(2)}$, ... and $\psi^{(1)}$, $\psi^{(2)}$, ..., etc., properly. Below we shall restrict ourselves to first order perturbation theory by neglecting all terms of the type $H^{(1)}\psi_k^{(1)}$, $H^{(2)}\psi_k^{(0)}$, etc., or of higher order. In this case Eq. 10.5 becomes

$$
H^{(0)}\psi_k^{(0)} + H^{(0)}\psi_k^{(1)} + H^{(1)}\psi_k^{(0)} = E_k^{(0)}\psi_k^{(0)} + E^{(0)}\psi_k^{(1)} + E^{(1)}\psi_k^{(0)} \quad (10.6)
$$

From Eq. 10.1 we know that $H^{(0)}\psi_k^{(0)} = E^{(0)}\psi_k^{(0)}$ so that Eq. 10.6 becomes

$$
H^{(0)}\psi_k^{1)} + H^{(1)}\psi_k^{(0)} = E^{(0)}\psi_k^{(1)} + E^{(1)}\psi_k^{(0)} \quad (10.6a)
$$

We assume that the eigenfunctions of $H^{(0)}$ form a complete set and expand $\psi_k^{(1)}$ in terms of this set:

$$
\psi_k^{(1)} = \sum_{m=1} a_{km}\psi_m^{(0)} \quad (10.7)
$$

Substituting this into Eq. 10.6a we obtain

$$
H^{(0)} \sum_{m=1} a_{km}\psi_m^{(0)} + H^{(1)}\psi_k^{(0)} = E_k^{(0)} \sum_{m=1} a_{km}\psi_m^{(0)} + E_k^{(1)}\psi_k^{(0)} \quad (10.8)
$$

If we multiply both sides of Eq. 10.8 from the left with $\psi_i^{(0)*}$ and integrate we obtain as a result of the orthonormality of the $\psi_i^{(0)}$

$$
E_i^{(0)}a_{ki} + H_{ik}^{(1)} = E_k^{(0)}a_{ki} + E_k^{(1)}\delta_{ki} \quad (10.9)
$$

From Eq. 10.9 we can obtain the a_{ki} and thus, by means of Eqs. 10.7 and 10.3, the perturbed eigenfunction $\psi_k = \psi_k^{(0)} + \psi_k^{(1)}$. First we assume that

$i = k$ and obtain from Eq. 10.9:

$$H_{kk}^{(1)} = E_k^{(1)} \qquad (10.10)$$

This does not tell us anything about the a_{ki} but states that *in first approximation the perturbation energy $E_k^{(1)}$ is equal to the expectation value of the perturbing Hamiltonian $H^{(1)}$ taken between the eigenfunctions $\psi_k^{(0)}$ of the unperturbed state.* For the case $i \neq k$ follows, if $E_k^{(0)}$ is nondegenerate (i.e., if $E_k^{(0)} \neq E_i^{(0)}$ for all $i \neq k$),

$$a_{ki} = \frac{H_{ik}^{(1)}}{E_k^{(0)} - E_i^{(0)}} \qquad (10.11)$$

This leaves only the coefficient a_{kk} undetermined. Its real part can be found to be zero from the normalization of $\psi_k = \psi_k^{(0)} + \psi_k^{(1)}$ (see Problem 10.2). The imaginary part of a_{kk} remains undetermined, and we can make it zero since it effects only the phase of ψ_k and has no influence on the energy of the perturbed state. To show this, we separate the term containing a_{kk} from the sum (Eq. 10.7):

$$\psi_k^{(1)} = \sum_{i=1} a_{ki}\psi_i^{(0)} = \sum_{i \neq k} a_{ki}\psi_i^{(0)} + a_{kk}\psi_k^{(0)}$$

Since we know that the coefficient a_{kk} is small as well as purely imaginary, we can write

$$\psi_k = (1 \pm i\,|a_{kk}|)\psi_k^{(0)} + \sum_{i \neq k} a_{ki}\psi_i^{(0)} \approx \exp(\pm i\,|a_{kk}|)\psi_k^{(0)} + \sum_{i \neq k} a_{ki}\psi_i^{(0)} \qquad (10.12)$$

This is the sum of a large $(\psi_k^{(0)})$ and a small $\left(\sum_{i \neq k} a_{ki}\psi_i^{(0)} \right)$ vector in the complex plane. To first order, the magnitude of this sum is clearly not affected if one of the two vectors is rotated by a small angle ($\varphi \approx |a_{kk}|$) with respect to the other.

With all the expansion coefficients thus determined, we can write down the perturbed eigenfunctions to first order:

$$\psi_k = \psi_k^{(0)} + \psi_k^{(1)} = \psi_k^{(0)} + \sum_{i \neq k} \frac{H_{ik}^{(1)}}{E_i - E_k} \psi_i^{(0)} \qquad (10.13)$$

This important equation tells us the following facts.

1. If the wave function of a state $\psi_k^{(0)}$ is changed by some perturbation $H^{(1)}$, we can express the perturbed wave function ψ_k as a linear combination of the unperturbed wave functions $\psi_i^{(0)}$. This is often expressed by saying that the perturbation *mixes* in other states.

2. The difference between the energy of the unperturbed state $\psi_k^{(0)}$ and the states $\psi_i^{(0)}$, which are mixed in by the perturbation, appears in the denominator of the

expansion coefficients a_{ki}. This means that those states whose energy E_i is close to the energy E_k of the unperturbed state are mixed in most heavily.

3. The matrix elements of the perturbation, $H_{ik}^{(1)} = (\psi_i^{(0)}, H^{(1)}\psi_k^{(0)})$ appear in the numerator of Eq. 10.13.

In view of the connection between the matrix elements and the transition probabilities discussed in Chapter 8.1 we can thus say: A perturbation $H^{(1)}$ mixes only those states between which a transition is possible under the influence of that perturbation.

By now it should have become obvious what the conditions are for the smallness of $\psi_k^{(1)}$ and $H^{(1)}$. If ψ_k is to differ only very little from $\psi_k^{(0)}$, only small amounts of the other states $\psi_i^{(0)}$ should be mixed in by the perturbation. In other words, the coefficients a_{ki} in Eq. 10.13 should be small. This, in turn, gives us the condition that qualifies $H^{(1)}$ as being small, namely,

$$1 \gg \left| \frac{H_{ik}}{E_i - E_k} \right| \qquad \text{for} \qquad i \gg k \qquad (10.14)$$

Thus far we have used the Schrödinger picture, but we can make the transition to matrix mechanics without difficulty. The perturbed eigenfunction ψ_k can be obtained from the unperturbed eigenfunctions $\psi_i^{(0)}$ through the "transformation"[3]:

$$\psi_k = \sum_{i=1} a_{ki}\psi_i^{(0)}, \qquad (a_{kk} = 1) \qquad (10.15)$$

According to Eq. 9.27, the same matrix A, whose elements transform the $\psi_i^{(0)}$ into the ψ_k, will transform the matrix $\mathbf{H}$ to its diagonal form through

$$\mathbf{A\dagger^*HA^* = E} \qquad (10.16)$$

10.2 THE PERTURBATION OF DEGENERATE STATES

The perturbation theory as we have developed it in the preceding paragraph has but one flaw: It is almost useless since it applies only to nondegenerate states. Not only are most atomic states degenerate, but the perturbation of degenerate states is of special interest to the experimental physicist. To understand this we repeat briefly the basic facts of degeneracy with special emphasis on the degeneracy of energy eigenvalues. An energy eigenvalue

---

[3] We have used the term transformation to emphasize the formal analogy to the transformation of vectors as described by Appendix A.4.5. We can, in this vein, consider the $\psi_i^{(0)}$ the elements of a *state vector*. The wary reader will sense yet another formulation of quantum mechanics lurking beyond the horizon.

$E^{(0)}$ is said to be n-fold degenerate if there are n linearly independent eigenfunctions $\psi_k^{(0)}$ so that

$$H^{(0)}\psi_k^{(0)} = E^{(0)}\psi_k^{(0)}, \qquad k = 1, 2, 3, \ldots, n \tag{10.17}$$

The fact that all the $\psi_k^{(0)}$ have the same energy eigenvalue $E^{(0)}$ with the Hamiltonian $H^{(0)}$ does not imply that they also have the same eigenvalue with some other Hamiltonian H. Even if H differs from $H^{(0)}$ only by a small perturbation,

$$H = H^{(0)} + H^{(1)} \tag{10.18}$$

If the perturbation $H^{(1)}$ acts differently on the various eigenfunctions $\psi_k^{(0)}$, it can very well happen that

$$H\psi_k = E_k\psi_k \tag{10.19}$$

where all the E_k are now *different*. In this case we say that *the degeneracy is lifted by the perturbation*. A degeneracy lifted by a perturbation is the delight of the experimental physicist because with luck and ingenuity it is often possible to measure directly the energy differences $E_k - E^{(0)}$ to a very high degree of accuracy (frequently much more accurately than the energy $E^{(0)}$ itself). In order to develop a perturbation theory applicable to degenerate states, we follow the approach taken in Chapter 10.1, amending it as needed.

Up to and including Eq. 10.6a, everything is all right; Eq. 10.10 on the other hand, cannot generally be true if the state $\psi_k^{(0)}$ is degenerate. In the case of n-fold degeneracy there exist n linearly independent eigenfunctions $\psi_{k_1}^{(0)}$, $\psi_{k_2}^{(0)}, \ldots, \psi_{k_n}^{(0)}$ and infinitely many linear combinations

$$\psi_k^{(0)} = \sum_{i=1}^{n} a_{k_i}\psi_k^{(0)} \tag{10.20}$$

all belonging to the same eigenvalue $E_k^{(0)}$. The perturbation, however, if it lifts the degeneracy (and this is the only case with which we shall concern ourselves), will do different things to these different unperturbed eigenfunctions. If we want to obtain the energy eigenvalues $E_{k_1}, \ldots, E_{k_n}$ that develop under the influence of the perturbation $H^{(1)}$, we have to pick those linear combinations $\psi_k^{(0)}$ that are eigenfunctions of $H^{(1)}$ or that, in matrix language, make $\mathbf{H}^{(1)}$ diagonal. That such linear combinations do exist can be seen in the following way. Assume that the perturbation has been applied and has lifted the degeneracy completely. In this case, n different nondegenerate eigenfunctions exist, each with one of the eigenvalues $E_{k_1}, \ldots, E_{k_n}$. If we now turn off the perturbation gradually, these n nondegenerate eigenfunctions will continuously go over into exactly n of the infinitely many linear combinations. These n linear combinations are sometimes called the adapted eigenfunctions of the perturbation. Once we have found the adapted eigenfunctions our troubles are over; we can then calculate the perturbation energy according to Eq. 10.10:

$$E_{kl} = (\psi_{kl}^{(0)}, H^{(1)}\psi_{kl}^{(0)}), \qquad l = 1, 2, \ldots, n \qquad (10.21)$$

At the same time Eq. 10.11 ceases to cause trouble. If $i \neq k$, the denominator still vanishes in the case of degeneracy, but so does the numerator ($H^{(1)}$ is diagonal), and we can hope that somehow we will be able to find the a_{ki}. Before we set out to do this, however, we convince ourselves in a more rigorous mathematical manner that the matrix $H^{(1)}$ can always be made diagonal in first approximation.

Let us assume the state $\psi_k^{(0)}$, in which we are interested, is 5-fold degenerate in the absence of a perturbation. In this case the diagonal energy matrix will look as follows.

$$\mathbf{H}^{(0)} = \begin{pmatrix} E_i^{(0)} & 0 & 0 & 0 & 0 & 0 & 0 & 0 & \cdots \\ 0 & E_j^{(0)} & 0 & 0 & 0 & 0 & 0 & 0 & \cdots \\ 0 & 0 & E_k^{(0)} & 0 & 0 & 0 & 0 & 0 & \cdots \\ 0 & 0 & 0 & E_k^{(0)} & 0 & 0 & 0 & 0 & \cdots \\ 0 & 0 & 0 & E & E_k^{(0)} & 0 & 0 & 0 & \cdots \\ 0 & 0 & 0 & 0 & 0 & E_k^{(0)} & 0 & 0 & \cdots \\ 0 & 0 & 0 & 0 & 0 & 0 & E_k^{(0)} & 0 & \cdots \\ 0 & 0 & 0 & 0 & 0 & 0 & 0 & E_l^{(0)} & \cdots \\ \cdot & \cdot & \cdot & \cdot & \cdot & \cdot & \cdot & \cdot \\ \cdot & \cdot & \cdot & \cdot & \cdot & \cdot & \cdot & \cdot \\ \cdot & \cdot & \cdot & \cdot & \cdot & \cdot & \cdot & \cdot \end{pmatrix} \qquad (10.22)$$

The energy eigenvalue $E_k^{(0)}$ is different from $E_i^{(0)}$, $E_j^{(0)}$, $E_l^{(0)}$, etc. Therefore these states will not give us any trouble in Eq. 10.12. The only states we have to worry about are the five states with $E_k^{(0)}$. This means, on the other hand, that the perturbation Hamiltonian $H^{(1)}$ has to be made diagonal only with respect to these five states. If $H^{(0)}$ and $H^{(1)}$ are to be simultaneously diagonal, (at least in the framed rows and columns) they must commute, (at least in the framed rows and columns). Now, the framed portion of $\mathbf{H}^{(0)}$ is proportional to the unit matrix, and the unit matrix commutes with any matrix; hence, we can always diagonalize $\mathbf{H}^{(1)}$ in the region of degeneracy.

In the general case of the perturbation of an n-fold degenerate state we therefore procede as follows.

(a) We pick n-different, linearly independent linear combinations $\psi_{k_1}^{(0)}$, $\psi_{k_2}^{(0)}$, $\ldots$, $\psi_{k_n}^{(0)}$ of the unperturbed eigenfunctions.

(b) We form the n^2 matrix elements of $\mathbf{H}^{(1)}$ with them.

(c) We diagonalize the resulting $n \times n$ matrix with a unitary transformation A (discussed in chapter 9.2.).

(d) If we are interested not only in the energy eigenvalues as given by the diagonalized form of $\mathbf{H}^{(1)}$, we can go on and find the adapted linear combinations by applying the unitary transformation A to

$$\psi_{k_1}^{(0)}, \psi_{k_2}^{(0)}, \ldots, \psi_{k_n}^{(0)}$$

Before we apply this technique to a specific example, we take a look at the shape of the matrices involved. The Hamiltonian matrix in the absence of a perturbation looks as follows

$$\mathbf{H}^{(0)} = \begin{pmatrix} E_1^{(0)} & 0 & 0 & 0 & 0 & 0 & 0 & 0 & 0 & \cdots \\ 0 & E_2^{(0)} & 0 & 0 & 0 & 0 & 0 & 0 & 0 & \cdots \\ 0 & 0 & E_3^{(0)} & 0 & 0 & 0 & 0 & 0 & 0 & \cdots \\ 0 & 0 & 0 & E_3^{(0)} & 0 & 0 & 0 & 0 & 0 & \cdots \\ 0 & 0 & 0 & 0 & E_3^{(0)} & 0 & 0 & 0 & 0 & \cdots \\ 0 & 0 & 0 & 0 & 0 & E_3^{(0)} & 0 & 0 & 0 & \cdots \\ 0 & 0 & 0 & 0 & 0 & 0 & E_3^{(0)} & 0 & 0 & \cdots \\ 0 & 0 & 0 & 0 & 0 & 0 & 0 & E_4^{(0)} & 0 & \cdots \\ 0 & 0 & 0 & 0 & 0 & 0 & 0 & 0 & E_5^{(0)} & \cdots \\ \cdot & \cdot & \cdot & \cdot & \cdot & \cdot & \cdot & \cdot & \cdot & \\ \cdot & \cdot & \cdot & \cdot & \cdot & \cdot & \cdot & \cdot & \cdot & \\ \cdot & \cdot & \cdot & \cdot & \cdot & \cdot & \cdot & \cdot & \cdot & \end{pmatrix}$$

$$(10.23)$$

Under the influence of the perturbation the Hamiltonian matrix, represented in the system of the unperturbed eigenfunctions $\psi_k^{(0)}$, becomes.

$$
\mathbf{H} =
\begin{pmatrix}
H_{11}^{(1)} + E_1^{(0)} & H_{12}^{(1)} & H_{13}^{(1)} & H_{14}^{(1)} & H_{15}^{(1)} & H_{16}^{(1)} & H_{17}^{(1)} & H_{18}^{(1)} & \cdots \\
H_{21}^{(1)} & H_{22}^{(1)} + E_2^{(0)} & H_{23}^{(1)} & H_{24}^{(1)} & H_{25}^{(1)} & H_{26}^{(1)} & H_{27}^{(1)} & H_{28}^{(1)} & \cdots \\
H_{31}^{(1)} & H_{32}^{(1)} & H_{33}^{(1)} + E_3^{(0)} & H_{34}^{(1)} & H_{35}^{(1)} & H_{36}^{(1)} & H_{37}^{(1)} & H_{38}^{(1)} & \cdots \\
H_{41}^{(1)} & H_{42}^{(1)} & H_{43}^{(1)} & H_{44}^{(1)} + E_3^{(0)} & H_{45}^{(1)} & H_{46}^{(1)} & H_{47}^{(1)} & H_{48}^{(1)} & \cdots \\
H_{51}^{(1)} & H_{52}^{(1)} & H_{53}^{(1)} & H_{54}^{(1)} & H_{55}^{(1)} + E_3^{(0)} & H_{56}^{(1)} & H_{57}^{(1)} & H_{58}^{(1)} & \cdots \\
H_{61}^{(1)} & H_{62}^{(1)} & H_{63}^{(1)} & H_{64}^{(1)} & H_{65}^{(1)} & H_{66}^{(1)} + E_3^{(0)} & H_{67}^{(1)} & H_{68}^{(1)} & \cdots \\
H_{71}^{(1)} & H_{72}^{(1)} & H_{73}^{(1)} & H_{74}^{(1)} & H_{75}^{(1)} & H_{76}^{(1)} & H_{77}^{(1)} + E_3^{(0)} & H_{78}^{(1)} & \cdots \\
H_{81}^{(1)} & H_{82}^{(1)} & H_{83}^{(1)} & H_{84}^{(1)} & H_{85}^{(1)} & H_{86}^{(1)} & H_{87}^{(1)} & H_{88}^{(1)} + E_4^{(0)} & \cdots \\
H_{91}^{(1)} & H_{92}^{(1)} & H_{93}^{(1)} & H_{94}^{(1)} & H_{95}^{(1)} & H_{96}^{(1)} & H_{97}^{(1)} & H_{98}^{(1)} & \cdots \\
\cdots & \cdots & \cdots & \cdots & \cdots & \cdots & \cdots & \cdots &
\end{pmatrix}
\tag{10.24}
$$

In constructing Eq. 10.24, we have assumed that arbitrary linear combinations of the degenerate eigenfunctions $\psi_k^{(0)}$ were used so that **H** is not in general diagonal. Now let us assume that we want to find out what happens to the state whose energy used to be $E_3^{(0)}$. The unitary matrix **A**, which transforms the region of interest into the diagonal form, looks as follows.

$$
\mathbf{A} = \begin{pmatrix}
1 & 0 & 0 & 0 & 0 & 0 & 0 & 0 & 0 & \cdots \\
0 & 1 & 0 & 0 & 0 & 0 & 0 & 0 & 0 & \cdots \\
0 & 0 & \boxed{\begin{matrix} a_{11} & a_{12} & a_{13} & a_{14} & a_{15} \\ a_{21} & a_{22} & a_{23} & a_{24} & a_{25} \\ a_{31} & a_{32} & a_{33} & a_{34} & a_{35} \\ a_{41} & a_{42} & a_{43} & a_{44} & a_{45} \\ a_{51} & a_{52} & a_{53} & a_{54} & a_{55} \end{matrix}} & 0 & 0 & \cdots \\
0 & 0 & 0 & 0 & 0 & 0 & 0 & 1 & 0 & \cdots \\
0 & 0 & 0 & 0 & 0 & 0 & 0 & 0 & 1 & \cdots \\
\vdots & & & & & & & & &
\end{pmatrix}
\tag{10.25}
$$

Obviously Eq. 10.25 will affect only the framed area of Eq. 10.24. As a result of the unitary transformation

$$A^{*\dagger}HA^* = H' \tag{10.26}$$

we thus get a Hamiltonian that is diagonal only in the area of interest (see page 176).

Equation 10.27 is the representation of the perturbed Hamiltonian in the system of the unperturbed eigenfunctions $\psi_k^{(0)}$ *if* instead of arbitrary linear combinations the adapted eigenfunctions have been used for the state with the energy $E_3^{(0)} + E_{3_k}^{(1)}$.

It should be emphasized that in order to obtain Eq. 10.27 and with it the desired values of the perturbation energies $E_{3_1}^{(1)} \cdots E_{3_5}^{(1)}$, we do not actually have to know the adapted eigenfunctions. Rather, by following the procedure of Chapter 9.2, we obtain $E_{3_1}^{(1)} \cdots E_{3_5}^{(1)}$ from the secular determinant (Eq. 9.36) and only then the elements of the unitary matrix A from the Eqs. 9.35.

Fortunately, we do not have to burden ourselves with those parts of Eqs. 10.24 and 10.25 that are outside the framed portions. We can write Eq. 10.26 just as well in the form

$$A^{*\dagger}HA^* = E^{(0)} + E^{(1)} \tag{10.28}$$

where $A^{*\dagger}$, H, A^*, $E^{(0)}$ and $E^{(1)}$ stand only for the framed portions of the

$$
\mathbf{H'} =
\begin{pmatrix}
H_{11}^{(1)} + E_1^{(0)} & H_{12}^{(1)} & H_{13}^{(1)} & H_{14}^{(1)} & H_{15}^{(1)} & H_{16}^{(1)} & H_{17}^{(1)} & H_{18}^{(1)} & \cdots \\
H_{21}^{(1)} & H_{22}^{(1)} + E_2^{(0)} & H_{23}^{(1)} & H_{24}^{(1)} & H_{25}^{(1)} & H_{26}^{(1)} & H_{27}^{(1)} & H_{28}^{(1)} & \cdots \\
H_{31}^{(1)} & H_{32}^{(1)} & E_3^{(0)} + E_{3_1}^{(1)} & 0 & 0 & 0 & 0 & H_{38}^{(1)} & \cdots \\
H_{41}^{(1)} & H_{42}^{(1)} & 0 & E_3^{(0)} + E_{3_2}^{(1)} & 0 & 0 & 0 & H_{48}^{(1)} & \cdots \\
H_{51}^{(1)} & H_{52}^{(1)} & 0 & 0 & E_3^{(0)} + E_{3_3}^{(1)} & 0 & 0 & H_{58}^{(1)} & \cdots \\
H_{61}^{(1)} & H_{62}^{(1)} & 0 & 0 & 0 & E_3^{(0)} + E_{3_4}^{(1)} & 0 & H_{68}^{(1)} & \cdots \\
H_{71}^{(1)} & H_{72}^{(1)} & 0 & 0 & 0 & 0 & E_3^{(0)} + E_{3_5}^{(1)} & H_{78}^{(1)} & \cdots \\
H_{81}^{(1)} & H_{82}^{(1)} & H_{83}^{(1)} & H_{84}^{(1)} & H_{85}^{(1)} & H_{86}^{(1)} & H_{87}^{(1)} & H_{88}^{(1)} + E_4^{(0)} & \cdots \\
\vdots & \vdots & \vdots & \vdots & \vdots & \vdots & \vdots & \vdots &
\end{pmatrix}
\tag{10.27}
$$

respective matrices. If we substitute $H = E^{(0)} + H^{(1)}$ into Eq. 10.28, we obtain

$$A^{*\dagger}(E^{(0)} + H^{(1)})A^* = E^{(0)} + A^{*\dagger}H^{(1)}A^* = E^{(0)} + E^{(1)} \quad (10.29)$$

or

$$A^{*\dagger}H^{(1)}A^* = E^{(1)} \quad (10.30)$$

This is an equation that directly relates the perturbation part of the Hamiltonian to the perturbation energy, and it is in this sense that we shall henceforth interpret the matrices A and $H^{(1)}$.

To become more familiar with this formalism we treat explicitly[4] a trivial case: the perturbation of the 2^1P_1 state of the He atom by a magnetic field. The total spin of this state is zero and so is the spin of the nucleus. Hence we are dealing with an orbital angular momentum of $l = 1$ and threefold degeneracy.

We have learned in Chapter 7.3 that a state with a total angular momentum[5] $\sqrt{l(l+1)}$ and a z-component m has a magnetic moment $\sqrt{l(l+1)}\mu_0$ with a z-component $m\mu_0$ where μ_0 is the Bohr magneton. We know from classical electrodynamics that the energy of a magnetic dipole with a dipole moment $\mathbf{\mu}$ in a magnetic field of field strength $\mathbf{B}$ is given by

$$E = (\mathbf{\mu} \cdot \mathbf{B}) \quad (10.31)$$

Since the magnetic moment of an electron is proportional to its angular momentum we assume, correctly but without proof, that the expectation value of its energy in a magnetic field is given by

$$(\psi, \mathbf{\mu} \cdot \mathbf{B}\psi) = (\psi, \mu_0\mathbf{L} \cdot \mathbf{B}\psi) = \mu_0(\psi,(L_xB_x + L_yB_y + L_zB_z)\psi) \quad (10.32)$$

The angular momentum operator $\mathbf{L}$ operates only on the angular part of the eigenfunctions ψ and leaves the radial eigenfunctions untouched. The angular parts of all eigenfunctions are the spherical harmonics.[6] The normalized spherical harmonics with $l = 1$ are

$$(a) \quad m = 1; \qquad Y_{11} = -\frac{1}{2}\sqrt{\frac{3}{2\pi}} \sin \vartheta e^{i\varphi}$$

$$(b) \quad m = 0; \qquad Y_{10} = \frac{1}{\sqrt{2}}\sqrt{\frac{3}{2\pi}} \cos \vartheta \qquad (10.33)$$

$$(c) \quad m = -1; \qquad Y_{1-1} = \frac{1}{2}\sqrt{\frac{3}{2\pi}} \sin \vartheta e^{-i\varphi}$$

For simplicity we assume the magnetic field to be in the z-direction, i.e.,

$$H^{(1)} = \mu_0(\mathbf{B} \cdot \mathbf{L}) = \mu_0 B_z L_z \quad (10.34)$$

---

[4] And in more detail than is actually necessary.

[5] In units of $\hbar$.

[6] As long as the potential depends only on r, which it does here. The second electron is in the $1s$ state and has, therefore, a spherically symmetrical distribution.

Since the three eigenfunctions are degenerate we try first to find the adapted linear combinations by forming $H_{ik}^{(1)}$, $i, k = 1, 0, -1$, with the eigenfunctions as given in Eq. 10.33a to c. According to Eq. 5.106 (letting $\hbar = 1$)

$$H^{(1)} = -i\mu_0 B_z \frac{\partial}{\partial \varphi} \tag{10.35}$$

Hence, $H^{(1)} Y_{10} = 0$ so that the middle column of our matrix $\mathbf{H}^{(1)}$ vanishes. The middle row must also be zero because Y_1 has no φ-dependent term and the term $e^{+i\varphi}$ integrated over 2π gives zero. Similarly,

$$(Y_{11}, H^{(1)} Y_{1-1}) \propto \int \int e^{-i\varphi} \sin \vartheta \frac{\partial}{\partial \varphi} (\sin \vartheta e^{-i\varphi}) \sin \vartheta \, d\vartheta \, d\varphi = 0 \tag{10.36}$$

because the integral over φ vanishes. The same holds for $(Y_{1-1}, H^{(1)} Y_{11})$. In other words, $\mathbf{H}^{(1)}$ *is* a diagonal matrix and Eq. 10.33a to c *are* the *adapted eigenfunctions*. If we calculate the two remaining matrix elements $(Y_{11}, H^{(1)} Y_{11})$ and $(Y_{1-1}, H^{(1)} Y_{1-1})$, we find them to be 1 and -1, respectively. The diagonal representation of the matrix $\mathbf{H}^{(1)}$ is therefore

$$\mathbf{H}^{(1)} = \mu_0 B_z \begin{pmatrix} 1 & 0 & 0 \\ 0 & 0 & 0 \\ 0 & 0 & -1 \end{pmatrix} \tag{10.37}$$

of course, this is no surprise; we could have read it directly from Eq. 10.34. The fact that the spherical harmonics are the adapted eigenfunctions is also trivial. They are, after all, eigenfunctions of the operator L_z.

We conclude from Eq. 10.37 that the singlet P_1 state splits into three levels under the influence of a magnetic field. The energy of the $m = 1$ level increases proportionally to the magnetic field; the energy of the $m = 0$ level remains unchanged; and the energy of the $m = -1$ level decreases proportionally to the magnetic field. If we plot the energy of each state as a function of the magnetic field, we obtain Figure 10.1. What if the magnetic field had been in the x-direction? The matrix would have been nondiagonal:

$$\mathbf{H}^{(1)} = \mu_0 B_x \mathbf{L}_x = \frac{\mu_0 B_x}{\sqrt{2}} \begin{pmatrix} 0 & 1 & 0 \\ 1 & 0 & 1 \\ 0 & 1 & 0 \end{pmatrix} \tag{10.38}$$

To diagonalize it, we have to solve according to Eq. 9.36:

$$\begin{vmatrix} -E_k & \dfrac{\mu_0 B_x}{\sqrt{2}} & 0 \\[2mm] \dfrac{\mu_0 B_x}{\sqrt{2}} & -E_k & \dfrac{\mu_0 B_x}{\sqrt{2}} \\[2mm] 0 & \dfrac{\mu_0 B_x}{\sqrt{2}} & -E_k \end{vmatrix} = 0 \tag{10.39}$$

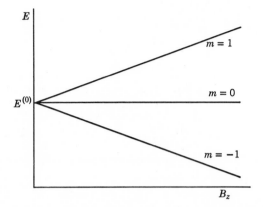

Fig. 10.1 An external magnetic field lifts the threefold degeneracy of an atom with $L = 1$.

This determinant has the solutions $E_k = 0$ and $E_k = \pm\mu_0 B_x$ as we would have expected since the energy of the atom should not depend on the direction of the magnetic field. Using Eq. 9.35, we determine the coefficients of the unitary transformation. Since

$$\frac{\mu_0 B_x}{\sqrt{2}} a_{2k}{}^* = a_{1k}{}^* E_k \tag{10.40}$$

$$\frac{\mu_0 B_x}{\sqrt{2}} a_{1k}{}^* + \frac{\mu_0 B_x}{\sqrt{2}} a_{3k}{}^* = a_{2k}{}^* E_k \tag{10.41}$$

$$\frac{\mu_0 B_x}{\sqrt{2}} a_{2k}{}^* = a_{3k}{}^* E_k \tag{10.42}$$

Letting $k = 1$ we find that, using $E_1 = \mu_0 B_x$,

$$a_{21}{}^* = \sqrt{2}\, a_{11}{}^* \tag{10.43}$$

and

$$a_{31}{}^* = a_{11}{}^* \tag{10.44}$$

For $k = 2$ it follows that (since $E_2 = 0$)

$$a_{22}{}^* = 0 \tag{10.45}$$

and

$$a_{12}{}^* = -a_{32}{}^* \tag{10.46}$$

Finally, with $k = 3$

$$a_{23}{}^* = -\sqrt{2}\, a_{13}{}^* \tag{10.47}$$

and

$$a_{13}{}^* = a_{33}{}^* \tag{10.48}$$

putting these equations together we get

$$\mathbf{A^*} = \begin{pmatrix} a_{11}{}^* & a_{12}{}^* & a_{13}{}^* \\ \sqrt{2}\,a_{11}{}^* & 0 & -\sqrt{2}\,a_{13}{}^* \\ a_{11}{}^* & -a_{12}{}^* & a_{13} \end{pmatrix} \tag{10.49}$$

The values of $a_{11}{}^*$, $a_{12}{}^*$, and $a_{13}{}^*$ follow from the unitarity of $\mathbf{A}$

$$\begin{pmatrix} a_{11}{}^* & a_{12}{}^* & a_{13}{}^* \\ \sqrt{2}\,a_{11}{}^* & 0 & -\sqrt{2}\,a_{13}{}^* \\ a_{11}{}^* & -a_{12}{}^* & a_{13} \end{pmatrix} \begin{pmatrix} a_{11} & \sqrt{2}\,a_{11} & a_{11} \\ a_{12} & 0 & -a_{12} \\ a_{13} & -\sqrt{2}\,a_{13} & a_{13} \end{pmatrix} = \begin{pmatrix} 1 & 0 & 0 \\ 0 & 1 & 0 \\ 0 & 0 & 1 \end{pmatrix} \tag{10.50}$$

Hence

$$a_{11}{}^*a_{11} + a_{12}{}^*a_{12} + a_{13}{}^*a_{13} = 1 \tag{10.51}$$

$$2a_{11}{}^*a_{11} + 2a_{13}{}^*a_{13} = 1 \tag{10.52}$$

From Eqs. 10.51 and 10.52 it follows that

$$a_{12}{}^*a_{12} = \tfrac{1}{2} \tag{10.53}$$

together with

$$\sqrt{2}\,a_{11}{}^*a_{11} - \sqrt{2}\,a_{13}{}^*a_{13} = 0 \tag{10.54}$$

and

$$a_{11}{}^*a_{11} - a_{12}{}^*a_{12} + a_{13}{}^*a_{13} = 0 \tag{10.55}$$

this yields

$$a_{11}{}^*a_{11} = a_{13}{}^*a_{13} = \tfrac{1}{4} \tag{10.56}$$

This does not say anything about the phase of the a_{ik}. However, since our one condition that $\mathbf{A}$ be unitary is satisfied whether $\mathbf{A}$ is real or not, we set the phase angle equal to zero and obtain

$$\mathbf{A} = \begin{pmatrix} \dfrac{1}{2} & \dfrac{1}{\sqrt{2}} & \dfrac{1}{2} \\[2mm] \dfrac{1}{\sqrt{2}} & 0 & -\dfrac{1}{\sqrt{2}} \\[2mm] \dfrac{1}{2} & -\dfrac{1}{\sqrt{2}} & \dfrac{1}{2} \end{pmatrix} \tag{10.57}$$

This enables us to find the adapted linear combinations

$$(a) \quad \psi_1 = \tfrac{1}{2}Y_1{}^1 + \frac{1}{\sqrt{2}}\,Y_1{}^0 + \tfrac{1}{2}Y_1{}^{-1}$$

$$(b) \quad \psi_2 = \frac{1}{\sqrt{2}}\,Y_1{}^1 - \frac{1}{\sqrt{2}}\,Y_1{}^{-1} \tag{10.58}$$

$$(c) \quad \psi_3 = \tfrac{1}{2}Y_1{}^1 - \frac{1}{\sqrt{2}}\,Y_1{}^0 + \tfrac{1}{2}Y_1{}^{-1}$$

To sum up: In the absence of a magnetic field, a He atom in the $2p$ state can be in any of the angular momentum states described by the eigenfunctions (Eqs. 10.32a to c) or any of their linear combinations. In other words, the angular momentum vector can point in any direction. In the presence of a magnetic field in the x-direction only the linear combinations (Eqs. 10.56a to c) are permitted. The atom precesses now in such a way that the x-component of the angular momentum vector has one of the values $\hbar$, 0, or $-\hbar$.[7]

10.3   SPIN, STATE VECTORS, HILBERT SPACE

We have stated earlier that spin is the intrinsic angular momentum of some elementary particles and that its existence cannot be deduced from a non-relativistic theory.

Since many interesting phenomena in atomic physics result from the existence of spin, we shall introduce spin, ad hoc, guessing at a formalism to deal with it. The formalism we shall use is due to **W. Pauli** and gives the essential features of the proper Dirac theory of the electron.

Formally we can introduce spin into the wave function of a quantum-mechanical system by adding two new quantum numbers: s for the magnitude of the spin angular momentum as given by[8]

$$\langle s^2 \rangle = s(s + 1) \tag{10.59a}$$

and m_s for its z-component as given by[8]

$$\langle s_z \rangle = m_s \tag{10.59b}$$

In order to avoid problems arising from the coupling of spin and orbital angular momentum, we assume the orbital angular momentum to be zero and write[9]

$$\psi = \psi_{n,s,m_s} \tag{10.60a}$$

Since spin is an intrinsic property of particles that exists regardless of their state of motion, we could also write ψ in the form

$$\psi = \psi_n \varphi_{s,m_s} \tag{10.60b}$$

where φ is the **spin eigenfuntion** and ψ_n the orbital eigenfunction of the system. Let us assume that Eq. 10.60 describes a system in which only one

---

[7] See Problem 10.8.

[8] In units of $\hbar^2$ or $\hbar$, respectively.

[9] The formalism we are about to develop can very well cope with the simultaneous existence of spin and orbital angular momentum; however, we shall restrict ourselves to the discussion of systems with $l = 0$ for simplicity.

particle has a spin and that this spin has $s = \frac{1}{2}$. What does the spin eigen-function look like in this case?

The coordinates of the spin eigenfunction φ_{s, m_s} cannot be the components x, y and z of the radius vector since the spatial distribution of the electron was supposed to be independent of its spin state. This means that, although spin constitutes an angular momentum in three-space, the coordinates of the spin eigenfunction refer to some other space which we shall call spin space.

We try to write the spin-eigenfunction in the form

$$\varphi_{m_s} = \frac{1}{\sqrt{2\pi}} e^{\pm i\gamma/2} \tag{10.61}$$

where φ is an angle in spin space. This seems to fit the bill. The functions φ_{m_s} are eigenfunctions of the operator[10]

$$L_z = i \frac{\partial}{\partial \gamma} \tag{10.62}$$

with eigenvalues $\pm\frac{1}{2}$. The functions φ_{m_s} are also orthonormal, since

$$\int_0^{2\pi} \varphi_{+\frac{1}{2}}^* \varphi_{-\frac{1}{2}} \, d\gamma = \frac{1}{2\pi} \int_0^{2\pi} e^{-i\gamma/2} e^{-i\gamma/2} \, d\gamma = 0, \tag{10.63a}$$

and

$$\int_0^{2\pi} \varphi_{\frac{1}{2}}^* \varphi_{\frac{1}{2}} \, d\gamma = 1 \tag{10.63b}$$

We can even form linear combinations

$$\varphi = a\varphi_{\frac{1}{2}} + b\varphi_{-\frac{1}{2}} \tag{10.64}$$

where $a^2 + b^2 = 1$ so that φ is normalized.

But that is as far as it goes. We cannot find any linear combinations that are eigenfunctions of L_x or L_y nor is the situation improved by the use of more complicated functions instead of $(1/\sqrt{2\pi})e^{\pm(i\gamma/2)}$ because such functions, if their linear combinations are eigenfunctions of L_x and L_y, make these operators nonhermitian. The plain fact is that φ_{m_s} is not a function in the usual sense, but is a **vector** in **spin space**.

It is usually written as

$$\varphi_{\frac{1}{2}} = \alpha = \begin{pmatrix} 1 \\ 0 \end{pmatrix} \tag{10.65}$$

if the *spin is up*, and

$$\varphi_{-\frac{1}{2}} = \beta = \begin{pmatrix} 0 \\ 1 \end{pmatrix} \tag{10.66}$$

[10] $\hbar = 1$.

if the *spin is down* and

$$\varphi = a\alpha + b\beta \tag{10.67}$$

where[11]

$$a^2 + b^2 = 1 \tag{10.68}$$

if it points somewhere else. α and β are **state vectors** in *spin space* and, according to the rules of vector algebra, they are orthonormal.

If our eigenfunctions are vectors, our operators must be matrices and

$$\mathbf{L}_z = \tfrac{1}{2}\sigma_z = \frac{1}{2}\begin{pmatrix} 1 & 0 \\ 0 & -1 \end{pmatrix} \tag{10.69}$$

obviously fits the bill for an operator that has α and β as eigenfunctions with the eigenvalues $+\tfrac{1}{2}$ and $-\tfrac{1}{2}$. α and β are not eigenfunctions of $\mathbf{L}_x$ and $\mathbf{L}_y$; instead we have

$$\mathbf{L}_x\alpha = \frac{1}{2}\begin{pmatrix} 0 & 1 \\ 1 & 0 \end{pmatrix}\begin{pmatrix} 1 \\ 0 \end{pmatrix} = \frac{1}{2}\begin{pmatrix} 0 \\ 1 \end{pmatrix} = \tfrac{1}{2}\beta \tag{10.70}$$

$$\mathbf{L}_x\beta = \frac{1}{2}\begin{pmatrix} 0 & 1 \\ 1 & 0 \end{pmatrix}\begin{pmatrix} 0 \\ 1 \end{pmatrix} = \frac{1}{2}\begin{pmatrix} 1 \\ 0 \end{pmatrix} = \tfrac{1}{2}\alpha \tag{10.71}$$

and

$$\mathbf{L}_y\alpha = \frac{1}{2}\begin{pmatrix} 0 & -i \\ i & 0 \end{pmatrix}\begin{pmatrix} 1 \\ 0 \end{pmatrix} = \frac{1}{2}\begin{pmatrix} 0 \\ i \end{pmatrix} = \frac{i}{2}\beta \tag{10.72}$$

$$\mathbf{L}_y\beta = \frac{1}{2}\begin{pmatrix} 0 & -i \\ i & 0 \end{pmatrix}\begin{pmatrix} 0 \\ 1 \end{pmatrix} = -\frac{1}{2}\begin{pmatrix} i \\ 0 \end{pmatrix} = -\frac{i}{2}\alpha \tag{10.73}$$

It is now possible to construct linear combinations of α and β that are normalized eigenfunctions (also called **eigenvectors**) of $\mathbf{L}_x$ and $\mathbf{L}_y$. For instance,

$$\mathbf{L}_x(a\alpha + b\beta) = \left(\frac{a}{2}\beta + \frac{b}{2}\alpha\right) = \tfrac{1}{2}(a\beta + b\alpha) \tag{10.74}$$

$(a\alpha + b\beta)$ is an eigenfunction of $\mathbf{L}_x$ if $a = b$. Its eigenvalue is $\tfrac{1}{2}$ and it can be normalized through $a = b = 1/\sqrt{2}$.

At this point some general remarks are in order. It is not just a stroke of luck that the multiplication of α and β with the angular momentum matrix

$$\mathbf{L}_z = \tfrac{1}{2}\sigma_z \tag{10.75}$$

yields the eigenvalue equations:

$$\mathbf{L}_z\alpha = \tfrac{1}{2}\alpha, \qquad \mathbf{L}_z\beta = -\tfrac{1}{2}\beta \tag{10.76}$$

There is more to it. We can obtain a consistent formulation of quantum mechanics that is completely equivalent to the Schrödinger (eigenfunction) or Heisenberg (matrix) formulation in the following way.

[11] Why?

A quantum mechanical system can be represented by a column vector that has one component for each of the possible combinations of all the quantum numbers that describe it. Since there can be infinitely many possible combinations, such **state vectors** can have infinitely many components. The infinite dimensional space spanned by these state vectors is called a **Hilbert space.**

Table 10.1.

Quantum numbers				State vectors			
n	1	j	m_j	(1)	(2)	(3)	(4)
1	0	$\frac{1}{2}$	$\frac{1}{2}$	a	0	0	0
1	0	$\frac{1}{2}$	$-\frac{1}{2}$	b	0	0	0
2	0	$\frac{1}{2}$	$\frac{1}{2}$	0	1	0	0
2	0	$\frac{1}{2}$	$-\frac{1}{2}$	0	0	0	0
2	1	$\frac{3}{2}$	$\frac{3}{2}$	0	0	a	0
2	1	$\frac{3}{2}$	$\frac{1}{2}$	0	0	b	0
2	1	$\frac{3}{2}$	$-\frac{1}{2}$	0	0	c	0
2	1	$\frac{3}{2}$	$-\frac{3}{2}$	0	0	d	0
2	1	$\frac{1}{2}$	$\frac{1}{2}$	0	0	0	1
2	1	$\frac{1}{2}$	$-\frac{1}{2}$	0	0	0	0
		etc.		0	0	0	0

Column 1 describes the atom in its $1S$ ground state in the absence of a magnetic field; the two spin states are degenerate; the normalization requires $a^2 + b^2 = 1$.

Column 2 describes a $2S$ state whose degeneracy has been lifted; the electron spin is up.

Column 3 describes a $2^2P_{3/2}$ state which is fourfold degenerate; the normalization requires $a^2 + b^2 + c^2 + d^2 = 1$.

Column 4 describes a $2^2P_{1/2}$ state whose degeneracy has been lifted; the electron spin is up.

A state vector is normalized to have the length 1. If the state of the system is described by a nondegenerate eigenfunction, the component of the state vector corresponding to its particular set of quantum numbers is 1. All others are zero. In the case of degeneracy several components can be non-zero; however, the sum of their squares has to be one.

As an example we write down state vectors for some states of the hydrogen atom in Table 10.1, disregarding the nuclear spin.

In the *Hilbert space picture* the place of the operators is taken by matrices and the place of the wave functions by vectors. If the multiplication of a vector with a matrix results in the same vector multiplied with a scalar, we call it an **eigenvector** of the matrix and the scalar its **eigenvalue.** All this has already been anticipated in our notation. The equation

$$(\psi, Q\psi) = \langle Q \rangle \tag{10.77}$$

can mean

$$\int \underbrace{\psi^* \overbrace{Q}^{\text{Operator}} \psi}_{\text{Wave functions}} \, d\tau = \overbrace{\langle Q \rangle}^{\text{Constant}} \tag{10.78}$$

or it can mean

$$(\psi_1{}^*, \psi_2{}^*, \psi_3{}^*, \ldots) \begin{pmatrix} q_{11} & q_{12} & q_{13} & \cdots \\ q_{21} & q_{22} & q_{23} & \cdots \\ q_{31} & q_{32} & q_{33} & \cdots \\ \cdot & \cdot & \cdot & \\ \cdot & \cdot & \cdot & \\ \cdot & \cdot & \cdot & \end{pmatrix} \begin{pmatrix} \psi_1 \\ \psi_2 \\ \psi_3 \\ \cdot \\ \cdot \\ \cdot \end{pmatrix} = \langle Q \rangle \tag{10.79}$$

Representative of state vector (row form) · matrix · Representative of state vector (column form) · Scalar

The result is, in either case, a number—the expectation value $\langle Q \rangle$. The row vector $\boldsymbol{\psi}\dagger = (\psi_1{}^*, \psi_2{}^*, \psi_3{}^*, \ldots)$ in Eq. 10.79 is the hermitian conjugate of the vector:

$$\psi = \begin{pmatrix} \psi_1 \\ \psi_2 \\ \cdot \\ \cdot \\ \cdot \end{pmatrix} \tag{10.80}$$

This enables us to use the matrix convention for the inner product (multiplication of every element in a row in the first matrix with the corresponding element in a column of the second matrix). The inner (dot) product of two vectors ($\boldsymbol{\psi} \cdot \boldsymbol{\phi}$) becomes thus $\boldsymbol{\psi}\dagger \cdot \boldsymbol{\phi}$. We do not want to pursue this matter further but simply state:

(a) Spin eigenfunctions can be written as two component vectors in spin space.

(b) The corresponding operators are the angular momentum matrices:[12]

$$\mathbf{L}_x = \tfrac{1}{2}\boldsymbol{\sigma}_x, \qquad \mathbf{L}_y = \tfrac{1}{2}\boldsymbol{\sigma}_y, \qquad \mathbf{L}_z = \tfrac{1}{2}\boldsymbol{\sigma}_z \tag{10.81}$$

where

$$\boldsymbol{\sigma}_x = \begin{pmatrix} 0 & 1 \\ 1 & 0 \end{pmatrix}, \qquad \boldsymbol{\sigma}_y = \begin{pmatrix} 0 & -i \\ i & 0 \end{pmatrix}, \qquad \boldsymbol{\sigma}_z = \begin{pmatrix} 1 & 0 \\ 0 & -1 \end{pmatrix} \tag{10.82}$$

are the Pauli matrices.

In the following paragraph we shall apply our knowledge of stationary perturbation theory and spin to the interaction between nuclear and electron spin.

10.4   HYPERFINE STRUCTURE

We have mentioned earlier that the proton *has spin* $\tfrac{1}{2}$.[13] Together with this intrinsic angular momentum goes a magnetic moment of 2.79275 nuclear magnetons.[14] The neutron also has spin $\tfrac{1}{2}$ and a magnetic moment of -1.9128 nuclear magnetons. The minus sign indicates that for equal direction of the spin angular momentum the magnetic moments of proton and neutron point in opposite directions. The fact that the neutron has a magnetic moment is somewhat surprising since very sensitive measurements have shown it to be strictly neutral. Its electric charge, if indeed it has any, is less than 10^{-20} electron charges. The existence of a magnetic moment shows that the neutron's disdain for electric fields is somewhat hypocritical as deep down inside it must have currents of equal but opposite charges. Since the proton and the neutron have a spin and a magnetic moment it is not surprising that most nuclei are similarly endowed. If the atomic electrons have a resulting angular momentum (and magnetic moment), the nuclear spin has to align itself in such a way that the resultant of nuclear spin and electron angular momentum

---

[12] $\hbar = 1$.

[13] This is the customary (though colloquial) way to express the fact that a particle has a spin quantum number $s = \tfrac{1}{2}$ and a total spin angular momentum $|L| = \sqrt{s(s+1)}\,\hbar = \sqrt{\tfrac{3}{4}}\,\hbar$.

[14] A nuclear magneton is a magnetic dipole moment $\mu_n = e\hbar/2m_p$, where m_p is the proton mass. Note the fact that the g-value of the proton is *not* 2 but 5.6. This large deviation from the Dirac theory is due to the nuclear forces of the proton.

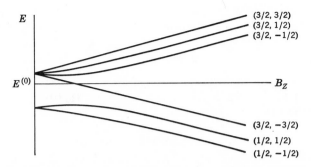

Fig. 10.2   The Zeeman effect of the hyperfine structure of deuterium.

can be written as[15]

$$\sqrt{f(f+1)}$$

according to the vector model (see Chapter 7.4). The various relative orientations of nuclear and electron magnetic moments have different energies. The electronic energy levels are thus split by the interaction of electronic and nuclear angular momentum or, in other words, the $2j + 1$ fold degeneracy of the electron state is partially lifted. Some examples may serve to illustrate this fact. The nucleus of the deuterium atom has spin 1. In the $1S$ ground state the (quantum number of the) electron angular momentum is $j = \frac{1}{2}$. The two possible hyperfine states have $f = \frac{1}{2}$ and $f = \frac{3}{2}$. One state is twofold degenerate, the other four-fold. Under the influence of an external magnetic field the state with $f = \frac{1}{2}$ will split into 2 levels and the state with $f = \frac{3}{2}$ into 4 levels (see Figure 10.2). In the $2^2P_{3/2}$ state of deuterium the total angular momentum of the electron is $j = \frac{3}{2}$. As a result we have three hyperfine levels with $f = \frac{1}{2}, f = \frac{3}{2}$ and $f = \frac{5}{2}$. The three hyperfine states are 2-, 4- and 6-fold degenerate.

The $2^2P_{1/2}$ state has $j = \frac{1}{2}$ and, therefore, $f = \frac{1}{2}$ and $f = \frac{3}{2}$ just as the ground state.

It should be pointed out that the total number of different possible orientations is not influenced by the coupling. Thus, in the P-state of deuterium we have one electron with $s = \frac{1}{2}$ and $l = 1$ and one nucleus with $I = 1$. That gives two possible orientations for the electron spin, three possible orientations for the electron orbital angular momentum, and three possible orientations for the nuclear spin, for a total of

$$2 \times 3 \times 3 = 18$$

different possibilities if there is no coupling.

[15] It is customary to describe the spin of the nucleus with I, m_I. The total angular momentum of the electron is usually described with j, m_j even if, as in an s-state, it consists only of the spin angular momentum. The resultant total angular momentum is described by f, m_f.

We must arrive at the same number if we add up all the degeneracies of the two P-states. The $2^2P_{1/2}$ state has one twofold and one fourfold degenerate level. The $2^2P_{3/2}$ state has three levels with 2-, 4-, and 6-fold degeneracy. Adding it up we get $2 + 4 + 2 + 4 + 6 = 18$.

10.5 THE HYPERFINE STRUCTURE OF THE HYDROGEN GROUND STATE

Before we discuss the hydrogen hyperfine structure quantitatively, we shall see what the vector model can tell us about it. Since the ground state is an s-state, we have $j = \frac{1}{2}$. The proton also has $I = \frac{1}{2}$, and proton and electron spin can thus be either *parallel* giving us a *triplet state* with $f = 1$, or they can be *antiparallel*, resulting in a *singlet state* with $f = 0$. The two states differ in energy, and this energy difference has been calculated to 5 or 6 decimal places. Experimentally it is one of the best known quantities in all of physics:

$$\nu = \frac{\Delta E}{h} = (1\ 420\ 405\ 751.800 \pm 0.028) \text{ cps} \qquad (10.83)$$

Spontaneous transitions between the two states are responsible for the famous 21 cm line emitted by interstellar hydrogen.

$$\frac{c}{\nu} = \lambda = 21 \text{ cm} \qquad (10.84)$$

The energy difference between the triplet and singlet state was first calculated in 1930 by E. Fermi who showed it to be

$$\Delta E = \hbar\omega = a\langle \mathbf{I} \cdot \mathbf{j} \rangle \qquad (10.85)$$

and calculated the value of the constant a. Fermi's result is what we might expect since the energy of two magnetic dipoles $\boldsymbol{\mu}_1$ and $\boldsymbol{\mu}_2$ is given by classical electrodynamics as

$$E \propto \boldsymbol{\mu}_1 \cdot \boldsymbol{\mu}_2 \qquad (10.86)$$

The rigorous derivation of Fermi's result (Eq. 10.85), however, goes beyond the scope of this book. Below we shall use the measured value of the **hyperfine splitting** between singlet and triplet state and derive from it the behavior of the two states in a magnetic field, i.e., the Zeeman effect. Our plan of attack will be the following.

First we shall consider the Coulomb interaction part of the Hamiltonian

$$H^{(0)} = -\frac{\hbar^2}{2m} \nabla^2 + V(r) \qquad (10.87)$$

as the Hamiltonian of the unperturbed atom. We shall add to this the perturbation due to the spin-spin interaction (Eq. 10.85)

$$H' = H^{(0)} + a(\mathbf{I} \cdot \mathbf{j}) \tag{10.88}$$

and find the adapted linear combinations of the unperturbed eigenfunctions (including, of course, the spin eigenfunctions). Next we shall consider the Hamiltonian (Eq. 10.88) as the unperturbed Hamiltonian whose energy eigenvalues (the *measured* hyperfine levels) and eigenfunctions (the adapted linear combinations) we know. We shall add to this Hamiltonian the perturbation due to the interaction with the external magnetic field

$$H = H' + \boldsymbol{\mu} \cdot \mathbf{B} \tag{10.89}$$

The spin-spin interaction did not lift the degeneracy completely; thus we again must form adapted linear combinations, this time starting with the eigenfunction that we found for the spin-spin interaction. So much for the plan, now to its implementation.

In the following we label all the quantities referring to the proton with a + sign and all the quantities referring to the electron with a − sign. The operator $(\mathbf{I} \cdot \mathbf{j})$ can be expressed in terms of the Pauli matrices

$$(\mathbf{I} \cdot \mathbf{j}) = \tfrac{1}{2}\boldsymbol{\sigma}^+ \cdot \tfrac{1}{2}\boldsymbol{\sigma}^- = \tfrac{1}{4}(\sigma_x{}^+\sigma_x{}^- + \sigma_y{}^+\sigma_y{}^- + \sigma_z{}^+\sigma_z{}^-) \tag{10.90}$$

where

$$\boldsymbol{\sigma}^+ = \begin{pmatrix} \sigma_x{}^+ \\ \sigma_y{}^+ \\ \sigma_z{}^+ \end{pmatrix} \tag{10.91}$$

acts only on the proton-spin eigenvector and

$$\boldsymbol{\sigma}^- = \begin{pmatrix} \sigma_x{}^- \\ \sigma_y{}^- \\ \sigma_z{}^- \end{pmatrix} \tag{10.92}$$

acts only on the electron-spin eigenvector. Using the symbols α and β for the spin eigenvector, we can write down the following combinations for the spins of electron and proton:

$$\varphi_1 = \alpha^+\alpha^- \qquad \varphi_2 = \beta^+\beta^- \tag{10.93}$$

(Both spins up)   (Both spins down)

(Notice that α^- and α^+ as well as β^- and β^+ are vectors in different spaces. The spin-space of the electron and the spin-space of the proton. $\alpha^+\alpha^-$, etc., are, therefore, not to be understood as inner products of the vectors α^+ and

α^-, etc., but merely as juxtapositions.)

$$\phi_3 = \alpha^+\beta^- \qquad \phi_4 = \beta^+\alpha^- \qquad (10.94)$$

(Proton spin up,   (Proton spin down,
electron spin down)   electron spin up)

From Eqs. 10.69 to 10.73 it follows that:

$$\sigma_x\alpha = \beta \qquad \sigma_x\beta = \alpha \qquad (10.95)$$

$$\sigma_y\alpha = i\beta, \qquad \sigma_y\beta = -i\alpha \qquad (10.96)$$

$$\sigma_z\alpha = \alpha, \qquad \sigma_z\beta = -\beta \qquad (10.97)$$

Substitution into Eq. 10.88 yields, using Eq. 10.90,[16]

$$a(\mathbf{I} \cdot \mathbf{j})\alpha^+\alpha^- = \frac{a}{4}(\sigma^+\sigma^-)\alpha^+\alpha^- = \frac{a}{4}(\sigma_x^+\sigma_x^- + \sigma_y^+\sigma_y^- + \sigma_z^+\sigma_z^-)\alpha^+\alpha^-$$

$$= \frac{a}{4}(\beta^+\beta^- - \beta^+\beta^- + \alpha^+\alpha^-) = \frac{a}{4}\alpha^+\alpha^- \qquad (10.98)$$

In other words $\alpha^+\alpha^-$ is one of the adapted linear combinations that make the matrix of the perturbing Hamiltonian diagonal. Next we try

$$\frac{a}{4}(\sigma^+ \cdot \sigma^-)\beta^+\beta^- = \frac{a}{4}(\alpha^+\alpha^- - \alpha^+\alpha^- + \beta^+\beta^-) = \frac{a}{4}\beta^+\beta^- \qquad (10.99)$$

Thus, $\beta^+\beta^-$ is also an eigenvector to $a(\mathbf{I} \cdot \mathbf{j})$; it is degenerate with $\alpha^+\alpha^-$. Now we try $\alpha^+\beta^-$

$$\frac{a}{4}(\sigma^+ \cdot \sigma^-)\alpha^+\beta^- = \frac{a}{4}(\beta^+\alpha^- + \beta^+\alpha^- - \alpha^+\beta^-) \qquad (10.100)$$

and $\beta^+\alpha^-$

$$\frac{a}{4}(\sigma^+ \cdot \sigma^-)\beta^+\alpha^- = \frac{a}{4}(\alpha^+\beta^- + \alpha^+\beta^- - \beta^+\alpha^-) \qquad (10.101)$$

Neither of these vectors is an eigenvector, and we should now proceed to use our diagonalization procedure; however, the adapted linear combinations are obvious in this case. We add Eq. 10.100 to Eq. 10.101:

$$\frac{a}{4}(\sigma^+ \cdot \sigma^-)(\alpha^+\beta^- + \beta^+\alpha^-) = \frac{a}{4}(\alpha^+\beta^- + \beta^+\alpha^-) \qquad (10.102)$$

This linear combination is an eigenvector, again with the eigenvalue $a/4$. We still have to normalize:

$$(\alpha^+\beta^- + \alpha^-\beta^+)(\alpha^+\beta^- + \alpha^-\beta^+) = 1 + 0 + 0 + 1 = 2 \qquad (10.103)$$

Hence

$$\psi_T = \frac{1}{\sqrt{2}}(\alpha^+\beta^- + \alpha^-\beta^+) \text{ is the normalized eigenvector.} \qquad (10.104)$$

[16] Remember that σ^+ operates only on α^+ and β^+ *not* on α^- or β^-; σ^- operates only on α^- and β^-.

If we subtract Eq. 10.101 from Eq. 10.100, we find another eigenvector:

$$\frac{a}{4}(\boldsymbol{\sigma}^+ \cdot \boldsymbol{\sigma}^-)(\alpha^+\beta^- - \beta^+\alpha^-) = \frac{a}{4}(3\beta^+\alpha^- - 3\alpha^+\beta^-) = -\frac{3a}{4}(\alpha^+\beta^- - \beta^+\alpha^-)$$

$$(10.105)$$

Its eigenvalue is $-\frac{3}{4}a$, and it can be normalized to give

$$\psi_S = \frac{1}{\sqrt{2}}(\alpha^+\beta^- - \beta^+\alpha^-) \qquad (10.106)$$

We sum up. The (normalized) adapted eigenvectors of the operator $a(\mathbf{I} \cdot \mathbf{j})$ of the spin-spin interaction are

$$\alpha^+\alpha^-, \beta^+\beta^- \quad \text{and} \quad \psi_T = \frac{1}{\sqrt{2}}(\alpha^+\beta^- + \beta^+\alpha^-) \quad (10.107)$$

They are degenerate and their energy is $a/4$ above the energy of the unperturbed ground state. There is also a nondegenerate state whose energy is $\frac{3}{4}a$ below the unperturbed level:

$$\psi_S = \frac{1}{\sqrt{2}}(\alpha^+\beta^- - \beta^+\alpha^-) \qquad (10.108)$$

The eigenvectors (Eq. 10.107) form the so-called *triplet state*, and ψ_S forms the *singlet state*. The energy difference between the two states is

$$\frac{a}{4} - \left(-\frac{3a}{4}\right) = a = \Delta E \qquad (10.109)$$

Now we determine the quantum numbers of the four adapted eigenfunctions. We start with m_f, the quantum number of the z-component of the total angular momentum.

Just from an inspection of the eigenfunctions we find that $\alpha^+\alpha^-$ has the eigenvalue $m_f = 1$ and $\beta^+\beta^-$ has the eigenvalue $m_f = -1$. Obviously both states have $f = 1$. Next we determine the eigenvalues of $(\alpha^+\beta^- \pm \beta^+\alpha^-)$

$$L_z \frac{1}{\sqrt{2}}(\alpha^+\beta^- \pm \beta^+\alpha^-) = \tfrac{1}{2}(\sigma_z^+ + \sigma_z^-)\frac{1}{\sqrt{2}}(\alpha^+\beta^- \pm \beta^+\alpha^-)$$

$$= \frac{1}{2\sqrt{2}}(\alpha^+\beta^- \mp \beta^+\alpha^- - \alpha^+\beta^- \pm \beta^+\alpha^-) \quad (10.110)$$

(i.e., both states have the eigenvalue $m_f = 0$.) Now we turn to the quantum number of the magnitude of the total angular momentum f. For $\alpha^+\alpha^-$ and $\beta^+\beta^-$ we have already found it. We can guess what it will be for the two remaining states, but it is instructive to carry out the calculation.

$$L^2 = (L^+ + L^-)^2 = \begin{pmatrix} L_x{}^+ + L_x{}^- \\ L_y{}^+ + L_y{}^- \\ L_z{}^+ + L_z{}^- \end{pmatrix}^2$$

$$= \tfrac{1}{4}(\sigma_x{}^{+2} + \sigma_x{}^{-2} + 2\sigma_x{}^+\sigma_x{}^- + \sigma_y{}^{+2} + \sigma_y{}^{-2} + 2\sigma_y{}^+\sigma_y{}^-$$
$$+ \sigma_z{}^{+2} + \sigma_z{}^{-2} + 2\sigma_z{}^+\sigma_z{}^-) \quad (10.111)$$

Now

$$\sigma_x{}^2 = \sigma_y{}^2 = \sigma_z{}^2 = 1 \quad (10.112)$$

Hence

$$L^2 \frac{1}{\sqrt{2}} (\alpha^+\beta^- \pm \beta^+\alpha^-)$$

$$= \frac{1}{4\sqrt{2}} [6 + 2(\sigma_x{}^+\sigma_x{}^- + \sigma_y{}^+\sigma_y{}^- + \sigma_z{}^+\sigma_z{}^-)](\alpha^+\beta^- \pm \beta^+\alpha^-)$$

$$= \frac{1}{4\sqrt{2}} [6(\alpha^+\beta^- \pm \beta^+\alpha^-) + 2(\beta^+\alpha^- \pm \alpha^+\beta^-) + 2(\beta^+\alpha^- \pm \alpha^+\beta^-)$$

$$-2(\alpha^+\beta^- \pm \beta^+\alpha^-)] = \frac{1}{\sqrt{2}} \begin{cases} \tfrac{8}{4}(\alpha^+\beta^- + \beta^+\alpha^-) \text{ selecting the upper sign} \\ 0(\alpha^+\beta^- - \beta^+\alpha^-) \text{ selecting the lower sign} \end{cases}$$

$$(10.113)$$

The eigenvalues of L^2 are thus 2 and 0, so that we have either $f = 1$ or $f = 0$. Table 10.2 sums it all up.

Table 10.2

State	Eigenfunction	Quantum numbers				Energy
		f	m_f	m_I	m_j	
	$\alpha^+\alpha^-$	1	1	$\tfrac{1}{2}$	$\tfrac{1}{2}$	$E_0 + \tfrac{1}{4}\Delta E$
Triplet	$\psi_T = \dfrac{1}{\sqrt{2}} (\alpha^+\beta^- + \beta^+\alpha^-)$	1	0	a	a	$E_0 + \tfrac{1}{4}\Delta E$
	$\beta^+\beta^-$	1	-1	$-\tfrac{1}{2}$	$-\tfrac{1}{2}$	$E_0 + \tfrac{1}{4}\Delta E$
Singlet	$\psi_S = \dfrac{1}{\sqrt{2}} (\alpha^+\beta^- - \beta^+\alpha^-)$	0	0	a	a	$E_0 - \tfrac{3}{4}\Delta E$

a m_I and m_j are not good quantum numbers for these states.

By now the reader may have wondered what became of the unperturbed Hamiltonian and the spatial part of the eigenfunctions. Nothing. The Hamiltonian $H^{(0)} = -(\hbar^2/2m)\nabla^2 + V(r)$ does not act on the spin eigenvectors (Eqs. 10.107 and 10.108), and the spin operator $\Delta E(\mathbf{I} \cdot \mathbf{j})$ leaves the spatial wave function untouched. In other words, the spatial part of the Hamiltonian matrix $\mathbf{H}^{(0)}$ is diagonal in the ψ_n and remains so regardless of the spin eigenvector with which we multiply ψ_n. All we have to do is, therefore, to diagonalize the perturbation and add the resulting terms to the already diagonal unperturbed matrix. The perturbed eigenfunctions are simply the product of the unperturbed spatial eigenfunctions and whatever spin eigenvectors we need to diagonalize the spin part of the Hamiltonian. After this preparation we are now ready to calculate the Zeeman Effect.

10.6 THE ZEEMAN EFFECT OF THE HYDROGEN HYPERFINE STRUCTURE

If we subject a hydrogen atom to an external magnetic field—assumed to be in the z-direction—the perturbing Hamiltonian becomes, according to Eqs. 10.88 and 10.89,

$$H = H^{(0)} + \Delta E(\mathbf{I} \cdot \mathbf{j}) + \boldsymbol{\mu} \cdot \mathbf{B} = H^{(0)} + \Delta E(\mathbf{I} \cdot \mathbf{j}) + \mu_z B_z \quad (10.114)$$

Here μ_z and B_z are the z-component of the magnetic dipole moment of the atom and the z-component of the external magnetic field. The magnetic moment of the hydrogen atom is the sum of the magnetic moments of proton and electron[17]

$$\mu_z = g^- \mu_0 \frac{\sigma_z^-}{2} - g^+ \mu_n \frac{\sigma_z^+}{2} = \mu^- \sigma_z^- - \mu^+ \sigma_z^+ \quad (10.115)$$

where μ_n is the nuclear magneton and μ_0 the Bohr magneton (g^+ and g^- are the g-values of the proton and electron). The minus sign takes into account that for equal spin-direction the magnetic moments of proton and electron have opposite directions because of the opposite signs of their charges. Thus, we can rewrite Eq. 10.114:

$$H = H^{(0)} + \frac{\Delta E}{4} (\boldsymbol{\sigma}^+ \cdot \boldsymbol{\sigma}^-) + B_z(\mu^- \sigma_z^- - \mu^+ \sigma_z^+) \quad (10.116)$$

To find the energy eigenvalues and the adapted eigenvectors of *this* perturbation we could, again, start with $\alpha^+\alpha^-$, $\alpha^+\beta^-$, $\beta^+\alpha^-$ and $\beta^+\beta^-$. But since we know that the eigenvectors of Eq. 10.116 must go over into the eigenvectors $\alpha^+\alpha^-$, ψ_T, $\beta^+\beta^-$ and ψ_S if the magnetic field is reduced to zero, it is more

---

[17] That is, in an s state. In the case of states with nonzero angular momentum there is also an orbital contribution.

instructive to express the eigenvectors of Eq. 10.116 as a linear combination of the latter.

We already know that $\alpha^+\alpha^-$, ψ_T, $\beta^+\beta^-$, and ψ_S are eigenvectors of the first two terms of Eq. 10.116, therefore, we investigate what happens if we operate on them with $B_z(\mu^-\sigma_z^- - \mu^+\sigma_z^+)$. We start with

$$B_z(\mu^-\sigma_z^- - \mu^+\sigma_z^+)\alpha^+\alpha^- = B_z(\mu^- - \mu^+)\alpha^+\alpha^- \qquad (10.117)$$

and

$$B_z(\mu^-\sigma_z^- - \mu^+\sigma_z^+)\beta^+\beta^- = -B_z(\mu^- - \mu^+)\beta^+\beta^- \qquad (10.118)$$

and find that both are eigenvectors of H, Eq. 10.116. The eigenvalues are $\pm B_z(\mu^- - \mu^+)$ so that the energy of these two hyperfine states in a magnetic field is

$$E_{\pm 1} = E^{(0)} + \frac{\Delta E}{4} \pm \frac{B_z}{2}(\mu_0 g^- - \mu_n g^+) \qquad (10.119)$$

Next we form

$$B_z(\mu^-\sigma_z^- - \mu^+\sigma_z^+)\frac{1}{\sqrt{2}}(\alpha^+\beta^- \pm \beta^+\alpha^-)$$

$$= \frac{B_z}{\sqrt{2}}(-\mu^-\alpha^+\beta^- \pm \mu^-\beta^+\alpha^- - \mu^+\alpha^+\beta^- \pm \mu^+\beta^+\alpha^-)$$

$$= -\frac{B_z}{\sqrt{2}}(\mu^- + \mu^+)(\alpha^+\beta^- \mp \beta^+\alpha^-) \qquad (10.120)$$

or, in other words,

$$B_z(\mu^-\sigma_z^- - \mu^+\sigma_z^+)\psi_S = -B_z(\mu^- + \mu^+)\psi_T \qquad (10.121)$$

and

$$B_z(\mu^-\sigma_z^- - \mu^+\sigma_z^+)\psi_T = -B_z(\mu^- + \mu^+)\psi_S \qquad (10.122)$$

This means that ψ_S and ψ_T are *not* eigenvectors of the perturbation by the external field. It is evident from Eqs. 10.121 and 10.122 that all the perturbation does is to mix these two states and that the adapted eigenvectors can be written as linear combinations of ψ_S and ψ_T. The simple sum $\psi_S + \psi_T$ does not suffice since, although it is an eigenvector of $B_z(\mu^-\sigma_z^- - \mu^+\sigma_z^+)$, it is not an eigenvector of Eqs. 10.116. We must then, alas, diagonalize Eq. 10.116, going by the book. According to Chapter 9.2, we start by writing "the matrix representation of the operator in some arbitrary complete set of functions." The obvious choice for these functions (here, of course, vectors) are the vectors ψ_S and ψ_T, which are eigenvectors in the absence of the magnetic field. Thus, we form

$$\mathbf{H}_{TT}' = (\psi_T, \{\Delta E(\mathbf{I} \cdot \mathbf{j}) + B_z(\mu^-\sigma_z^- - \mu^+\sigma_z^+)\}\psi_T), \quad \text{etc.} \quad (10.123)$$

According to Eqs. 10.102 and 10.122, we have[18]

$$(\psi_T, \Delta E(\mathbf{I} \cdot \mathbf{j})\psi_T) = \left(\psi_T, \frac{\Delta E}{4}(\sigma^+ \cdot \sigma^-)\psi_T\right) = \frac{\Delta E}{4} \qquad (10.124)$$

and

$$(\psi_T, B_z(\mu^- \sigma_z^- - \mu^+ \sigma_z^+)\psi_T) = 0 \qquad (10.125)$$

Using Eqs. 10.105 and 10.121, we can obtain the three other matrix elements and find the following representation:

$$H^{(1)\prime} = \begin{pmatrix} \dfrac{\Delta E}{4} & -B_z(\mu^- + \mu^+) \\ -B_z(\mu^- + \mu^+) & -\dfrac{3\Delta E}{4} \end{pmatrix} \qquad (10.126)$$

The next step is to "find the unitary transformation that transforms this matrix into a diagonal matrix."[19] According to Eq. 9.33, this unitary transformation $\mathbf{A}$ is defined by

$$\mathbf{H'A^*} = \mathbf{A^*E} \qquad (10.127)$$

or written explicitly for our two-by-two matrix:

$$\begin{pmatrix} \dfrac{\Delta E}{4} & -B_z(\mu^- + \mu^+) \\ -B_z(\mu^- + \mu^+) & -\dfrac{3\Delta E}{4} \end{pmatrix} \begin{pmatrix} a_{11}{}^* & a_{12}{}^* \\ a_{21}{}^* & a_{22}{}^* \end{pmatrix}$$

$$= \begin{pmatrix} a_{11}{}^* & a_{12}{}^* \\ a_{21}{}^* & a_{22}{}^* \end{pmatrix} \begin{pmatrix} E_1 & 0 \\ 0 & E_2 \end{pmatrix} \qquad (10.128)$$

This matrix equation yields four linear homogeneous equations for the four coefficients $a_{ik}{}^*$. "These linear homogeneous equations can be solved if, and only if, the secular determinant vanishes."[20]

$$\begin{vmatrix} \dfrac{\Delta E}{4} - E & -B_z(\mu^- + \mu^+) \\ -B_z(\mu^- + \mu^+) & -\dfrac{3\Delta E}{4} - E \end{vmatrix} = 0 \qquad (10.129)$$

This is a quadratic equation

$$E^2 - \tfrac{3}{16}\Delta E^2 + \frac{E\,\Delta E}{2} - B_z{}^2(\mu^- + \mu^+)^2 = 0 \qquad (10.130)$$

[18] Remember, $a = \Delta E$.
[19] See Chapter 9.2.
[20] See Problem 10.7.

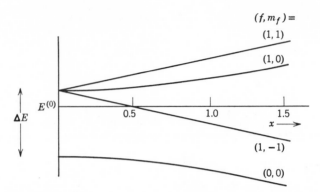

Fig. 10.3 The Breit-Rabi diagram of the hyperfine structure of the hydrogen ground state.

which has the solutions:

$$E = -\frac{\Delta E}{4} \pm \sqrt{\frac{\Delta E^2}{4} + B_z^2(\mu^- + \mu^+)^2}$$ (10.131)

Using the abbreviation

$$\frac{2B_z}{\Delta E}(\mu^- + \mu^+) = \frac{B_z}{\Delta E}(\mu^0 g^- + \mu_n g^+) = x$$ (10.132)

Eq. 10.131 is often written in the form:

$$E = -\frac{\Delta E}{4} \pm \frac{\Delta E}{2}\sqrt{1 + x^2}$$ (10.133)

Adding to this the energy $E^{(0)}$ of the unperturbed level, we obtain for the energy of the states ψ_T and ψ_S, respectively,

$$E_{T,S} = E^{(0)} - \frac{\Delta E}{4} \pm \frac{\Delta E}{2}\sqrt{1 + x^2}$$ (10.134)

Equation 10.134 is a special case of the Breit-Rabi formula[21] which describes the Zeeman effect of the atomic hyperfine structure.

With Eqs. 10.119 and 10.134, we have completely unravelled the hyperfine structure of the hydrogen ground state. Figure 10.3 shows a plot of the Eqs. 10.119 and 10.134, the so called Breit-Rabi diagram. We could now easily calculate the coefficients of the unitary transformation $\mathbf{A}^*$, but we postpone this until we have some practical application for them.

PROBLEMS

10.1 In analogy to Eq. 10.6a derive from Eq. 10.5 an expression that includes terms small of second order.

[21] G. Breit and I. I. Rabi, Phys. Rev. **38**, 2082 (1931).

10.2 Show that to *first order* the real part of the expansion coefficients a_{kk} vanish if we assume that both $\psi_k^{(0)}$ and $\psi_k = \psi_k^{(0)} + \psi_k^{(1)}$ are normalized:

$$(\psi_k, \psi_k) = (\psi_k^{(0)}, \psi_k^{(0)}) = 1$$

10.3 Show that from

$$Q' = A^*QA\dagger^*$$

follows

$$Q = A\dagger^*Q'A^*$$

if A is a unitary matrix.

10.4 Find linear combinations of α and β that are eigenvectors of L_y and L^2.

10.5 Show that $\alpha^+\alpha^-$, $\beta^+\beta^-$, $\psi_T = (1/\sqrt{2})(\alpha^+\beta^- + \beta^+\alpha^-)$ and $\psi_S = (1/\sqrt{2})(\alpha^+\beta^- - \beta^+\alpha^-)$ are mutually orthogonal.

10.6 Verify Eq. 10.126.

10.7 Equation 10.128 yields four equations. Why don't we solve the secular determinant for the second pair of equations?

10.8 Muonium is an atom consisting of an electron and a positive muon. The muon has spin $\frac{1}{2}$. Its mass is 206 times the electron mass. The muon behaves in all respects like a heavy electron. In particular, its magnetic moment is very close to μ_μ, the magnetic moment we obtain if we replace the electron mass with the muon mass in the expression for the Bohr magneton μ_0. The energy difference between the states ψ_T and ψ_S in the absence of a magnetic field is

$$\Delta E = h\nu = h(4463.15 \pm 0.06 \text{ Mc/sec})$$

In a magnetic field the energy difference between the states $(f, m_f) = (1, 1)$ and $(f, m_f) = (1, 0)$ was measured and found to be $h \cdot (1800 \text{ Mc/sec})$. What was the magnetic field strength?

10.9 We can calculate the energy difference between the states ψ_T and ψ_S of a muonium atom in the *absence* of a magnetic field from a measurement of the energy difference between the states $(f, m_f) = (1, 1)$ and $(f, m_f) = (1, 0)$ (see previous problem) in the *presence* of an external magnetic field. In order to minimize the error resulting from the inhomogeneity of the field, it would be desirable to make the measurement at a value of the magnetic field at which the energy difference is field-independent. Does such a value exist? Give a qualitative argument. If such a value of the magnetic field exists, calculate it.

10.10 For very large values of the magnetic field B_z, do the energies of the states $\alpha^+\alpha^-$ and ψ_T converge, diverge, or approach a constant difference?

10.11 At what value of the external magnetic field B_z is the energy difference between the states $\beta^+\beta^-$ and ψ_T equal to the energy difference ΔE between the unperturbed states ψ_T and ψ_S?

10.12 At what value (other than zero) of the external magnetic field B_z is the energy difference between the states $\beta^+\beta^-$ and ψ_S equal to ΔE, the energy difference between the unperturbed state ψ_T and ψ_S?

SOLUTIONS

10.2 We have

$$1 = (\psi_k^{(0)}, \psi_k^{(0)}) = (\psi_k, \psi_k) = (\psi_k^{(0)} + \psi_k^{(1)}, \psi_k^{(0)} + \psi_k^{(1)})$$
$$= (\psi_k^{(0)}, \psi_k^{(0)}) + (\psi_k^{(0)}, \psi_k^{(1)}) + (\psi_k^{(1)}, \psi_k^{(0)}) + (\psi_k^{(1)}, \psi_k^{(1)})$$

In first approximation we have

$$(\psi_k^{(1)}, \psi_k^{(1)}) \approx 0$$

Hence, since $\psi_k^{(0)}$ was assumed to be normalized,

$$(\psi_k^{(0)}, \psi_k^{(1)}) = -(\psi_k^{(1)}, \psi_k^{(0)}) = -(\psi_k^{(0)}, \psi_k^{(1)})^*$$

In other words, the matrix element $(\psi_k^{(0)}, \psi_k^{(1)})$ is imaginary. Now we multiply Eq. 10.7 with $\psi_k^{(0)*}$. Since the $\psi_k^{(0)}$ are orthonormal we get

$$(\psi_k^{(0)}, \psi_k^{(1)}) = a_{kk}(\psi_k^{(0)}, \psi_k^{(0)}) = a_{kk}$$

Since $(\psi_k^{(0)}, \psi_k^{(1)})$ was shown to be imaginary the a_{kk} do not have a real part.

10.12 The energy of the state $\beta^+\beta^-$ is according to Eq. 10.119

$$E_{-1} = E^{(0)} + \frac{\Delta E}{4} - \frac{B_z}{2}(\mu_o g^- - \mu_n g^+)$$

The energy of the state ψ_S is given by Eq. 10.134 as

$$E_s = E^{(0)} - \frac{\Delta E}{4} - \frac{\Delta E}{2}\sqrt{1 + x^2}$$

Thus the energy difference between the two states is

$$E_{-1} - E_s = \frac{\Delta E}{2} + \frac{\Delta E}{2}\sqrt{1 + x^2} - \frac{B_z}{2}(\mu_o g^- - \mu_n g^+)$$

Setting this equal to ΔE we obtain

$$\frac{\Delta E}{2}(1 - \sqrt{1 + x^2}) + \frac{B_z}{2}(\mu_o g^- - \mu_n g^+) = 0$$

Substituting the proper value (Eq. 10.132) for x, we obtain

$$\frac{\Delta E}{2}\left(1 - \sqrt{1 + \frac{B_z^2}{\Delta E^2}(\mu_o g^- + \mu_n g^+)^2}\right) + \frac{B_z}{2}(\mu_o g^- - \mu_n g^+) = 0$$

Solving this for B_z, we obtain

$$B_z = \frac{(\mu_o g^- - \mu_n g^+) \Delta E}{2\mu_o g^- \cdot \mu_n g^+}$$

Now $\mu_o/g^- = \mu_e = 0.92838 \cdot 10^{-20}$ erg/gauss is the magnetic moment of the electron, and $\mu_n g^+ = \mu_p = 14.105 \cdot 10^{-23}$ erg/gauss that of the proton. Since $\mu_p \ll \mu_e$, we can neglect μ_p in the numerator so that (using Eq. 10.83)

$$B_z \approx \frac{\Delta E}{2\mu_p} = \frac{h\nu}{2\mu_p} = \frac{6.626 \cdot 10^{-27} \text{ erg/sec } 1.42 \cdot 10^9 \text{ sec}^{-1}}{2 \cdot 1.41 \cdot 10^{-23} \text{ erg gauss}^{-1}} = 3.34 \cdot 10^5 \text{ gauss}$$

11

POSITRONIUM

11.1 A LOWBROW APPROACH

If a positron is brought to rest in matter, it ionizes the material along its path. Toward the end of its trail, when its velocity is sufficiently reduced, it can pick up an electron and form a positronium atom. If the process happens in a noble gas, the positronium atom should live happily ever after as a free atom. But how long is ever after? Since

electron + anti-electron = electron − electron = no electron

a positronium atom lives on borrowed time, and reality sooner or later will catch up with it, that is, positron and electron will annihilate each other. Where does the energy 2 mc² go? Because of its low mass there are only two particles into which the positronium atom can decay—the neutrino and the photon. Both lead a very hurried existence, always moving with the velocity of light and, thus, both have a nonzero momentum in any coordinate system.

If the positronium atom is at rest (and in its own center-of-mass coordinate system it is always at rest), decay into one moving particle does not conserve momentum. Positronium, therefore, has to decay into at least two particles.

A decay into neutrinos, although possible, involves the so-called weak interaction and is, therefore, vastly less probable than the decay into photons, which proceeds by means of the much stronger electromagnetic interaction.

A detailed analysis (see Chapter 11.4) shows that positronium in its singlet state can decay into 2 photons. To conserve momentum and energy, these two photons must have precisely the same energy, and they must be emitted at a relative angle of precisely 180° if the positronium atom was at rest at the moment of its demise. The energy of each photon must be

$$E = \mathrm{mc^2} = 511 \text{ keV}$$

the rest energy of the electron. Positronium in its triplet state can only decay into an odd number of photons, and the smallest odd number (other than one) is three. Although decays into five or seven photons are possible, they are less probable and have not been observed. There are infinitely many ways to split energy and momentum between three photons, and no simple rules exist as to their angular distribution and energy distribution except that all three have to be emitted in one plane.

The singlet state has a lifetime of 1.25×10^{-10} seconds,[1] and the triplet state has a lifetime of 1.4×10^{-7} seconds. The existence of this long-lived component is an unmistakable sign of the formation of positronium.

11.2   THE HYPERFINE-STRUCTURE OF POSITRONIUM

In Chapter 5.7 we discussed the energy-level structure of positronium and we have nothing to add except that none of the predicted spectral lines have ever been observed. This is not disturbing since positronium is a very rare commodity and since it is usually formed in its ground state. However, since positrons have spin $\frac{1}{2}$ the positronium atom has a hyperfine structure, and the hyperfine structure of its $1s$ ground state has been measured by Deutsch and Brown in a very elegant experiment, which we shall discuss later. First, we shall see what we can find out theoretically about the hyperfine structure of positronium.

As in the case of hydrogen, the ground state must be split by the $\mathbf{I} \cdot \mathbf{j}$ interaction into a singlet and a triplet state. Also, as in the case of hydrogen, *we* cannot calculate this splitting not knowing Dirac-theory, let alone quantum electrodynamics. The measured energy difference between the two states is

$$E = h\nu = h(2.0337 \times 10^5 \text{ Mc/sec}) \qquad (11.1)$$

in perfect agreement with the theoretical prediction. This value is much larger than that of the hyperfine splitting of hydrogen (as well it should be), since the positron has a much larger magnetic moment[2] than the proton. Having been given the value of the hyperfine splitting, we can now proceed to calculate the Zeeman effect of the positronium ground state.

Apart from the magnetic moment of the positive particle, nothing has changed when compared with the case of the hydrogen atom discussed above. $\alpha^+\alpha^-$ and $\beta^+\beta^-$ are still adapted eigenfunctions with the eigenvalues

$$\pm B_z(\mu^- - \mu^+)$$

But this time we have

$$\mu^- = \mu^+$$

---

[1] In metals, positrons annihilate with the conduction electrons without forming positronium. The annihilation lifetime in this case is also of the order of 10^{-10} seconds.

[2] The magnetic moment of the positron is, of course (except for its sign), the same as that of the electron.

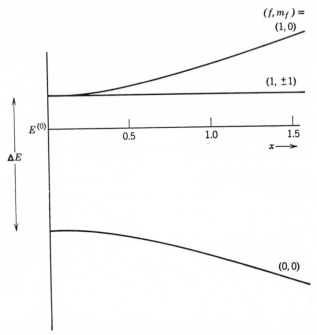

Fig. 11.1 The Breit-Rabi diagram of the hyperfine structure of the positronium ground state.

so that both eigenvalues vanish. This means that the external magnetic field does not change the energy levels of the states $\alpha^+\alpha^-$ and $\beta^+\beta^-$. The states ψ_T or $(f, m_f) = (1, 0)$ and ψ_S or $(f, m_f) = (0, 0)$ are still described by the Breit-Rabi formula (Eq. 10.131), but this time $(\mu^- + \mu^+)$ enters and there is only a quantitative change in the behavior of the two levels in the magnetic field. The resulting Breit-Rabi diagram of the ground state of positronium is shown in Figure 11.1.

11.3 THE MIXING OF STATES

We have learned in Chapter 10.1 that a perturbation can mix states. In other words, a system, that (in the absence of a perturbation) is in a state ψ_i goes under the influence of a perturbation into another state φ_i. The eigen-function φ_i of the complete Hamiltonian (including the perturbation) can be written as a superposition of possible eigenfunctions of the unperturbed system,

$$\varphi_i = \sum_k c_{ik}\psi_k \qquad (11.2)$$

where the expansion coefficients c_{ik} can be determined by the methods developed in Chapter 10. Equation 11.2 can be interpreted as meaning that the originally "pure state" ψ_i has been "mixed" with some or all of the other eigenstates ψ_k of the unperturbed Hamiltonian. The degree of the mixing is given by the coefficients c_{ik}, and in Eqs. 9.35 and 10.11 we have been given prescriptions for finding them. We have seen, however, that the energy eigenvalues of the problem can be obtained without worrying about the actual values of the coefficients. This raises the following question: Is the superposition of states just a mathematical exercise designed to gain useless information, or is the mixing an observable physical phenomenon? The answer is, the latter, as shown by the following example.

The singlet 1-s state of positronium has a lifetime of 1.25×10^{-10} sec., whereas the triplet state has a lifetime of 1.4×10^{-7} sec. If we mix the two states, the lifetime of the mixed state should be somewhere in between and we should be able to calculate precisely where in between.[3]

Before we undertake a detailed calculation we survey the situation qualitatively:

The $\alpha^+\alpha^-$ and $\beta^+\beta^-$ components of the triplet state are adapted to the perturbation; they are not mixed with anything and will maintain their lifetime of 1.4×10^{-7} sec.

ψ_T and ψ_S will be mixed by the magnetic field. This means that an atom originally in the ψ_S state now has a chance to decay slowly via three quantum emission. Since it still has the chance to decay rapidly by way of two quantum emission, its lifetime will remain of the order 10^{-10} sec.

For an atom originally in the ψ_T state the situation is radically different. The admixture of even a small amount of ψ_S opens the two quantum decay "channel" in which decays are a thousand times more likely. We can, therefore, expect that the lifetime of the ψ_T component is roughly halved if we mix in as little as 0.1 percent ψ_S. Before we go any further we must ask: What is the probability that the atom is in any particular state to begin with? Let us assume that neither the electrons nor the positrons used in the manufacture of positronium are polarized. This means that there are as many electrons and positrons in the α state as there are in the β state. The probability that an electron is in the α state is thus $\frac{1}{2}$; the probability that this electron captures a positron in the α state is also $\frac{1}{2}$. Therefore, we expect to find one quarter of all positronium atoms in the $\alpha^+\alpha^-$ state. The same arguments apply to the $\beta^+\beta^-$ state which therefore is also populated by one quarter of the atoms. The states $\beta^+\alpha^-$ and $\alpha^+\beta^-$ are not eigenstates but are mixed by the $(\mathbf{I} \cdot \mathbf{j})$ interaction. Both have a probability of one quarter to occur and both are equally represented in $\psi_S = (1/\sqrt{2})(\alpha^+\beta^- - \beta^+\alpha^-)$ and

---

[3] Implicit in this statement is the assumption that the lifetime of the pure states is not changed by the magnetic field, a reasonable assumption, since the spatial distribution of the electron and positron which determines the lifetimes together with the relative spin orientation, remains unchanged.

$\psi_T = (1/\sqrt{2})(\alpha^+\beta^- + \beta^+\alpha^-)$. Therefore, both these states also occur with a probability of 25 percent. Thus, in the absence of a magnetic field we have the following situation: Three quarters of all the positronium atoms are in the threefold degenerate triplet $1s$ state and one quarter are in the singlet $1s$ state. As a consequence 75 percent of the atoms will decay into three quanta with a lifetime of 1.4×10^{-7} sec and 25 percent will decay into two quanta with a lifetime of 1.25×10^{-10} sec.

If we turn on a magnetic field the situation changes: One half of the atoms will continue to decay with the appropriate lifetime into three quanta. One quarter will decay into two quanta with a lifetime of the order of 10^{-10} sec., and one quarter will decay into two quanta with a field-dependent lifetime *somewhere* in between. Now, we shall try to find out just where *somewhere* is.

The matrix equation (Eq. 10.128) yields four linear homogeneous equations for the a_{ik}^*. The first one is (using Eqs. 10.132 and 10.133)

$$a_{11}^* \frac{\Delta E}{4} - a_{21}^* \frac{\Delta E}{2} x = a_{11}^* \left(-\frac{\Delta E}{4} + \frac{\Delta E}{2} \sqrt{1 + x^2} \right) \tag{11.3}$$

or

$$\frac{a_{11}^*}{a_{21}^*} = \frac{x}{1 - \sqrt{1 + x^2}} = \gamma \tag{11.4}$$

Similarly we get

$$\frac{a_{12}^*}{a_{22}^*} = \frac{\sqrt{1 + x^2} - 1}{x} = \frac{-1}{\gamma} \tag{11.5}$$

The new eigenvectors we are after are (according to Eq. 9.17)

$$\psi_1 = a_{11}\psi_T + a_{12}\psi_S \tag{11.6}$$

and

$$\psi_2 = a_{12}\psi_T + a_{22}\psi_S \tag{11.7}$$

To determine the real and the imaginary parts of the four complex coefficients a_{ik}^*, we need eight equations, two of which we have already found (Eqs. 11.4 and 11.5). The unitarity of A^* provides four more equations

$$\begin{pmatrix} a_{11}^* & a_{12}^* \\ a_{21}^* & a_{22}^* \end{pmatrix} \begin{pmatrix} a_{11}^* & a_{21} \\ a_{12} & a_{22}^* \end{pmatrix} = \begin{pmatrix} 1 & 0 \\ 0 & 1 \end{pmatrix} \tag{11.8}$$

but there it ends. We are thus forced to economize somewhere, and a look at Eqs. 11.6 and 11.7 suggests just the place. The absolute value of the phase of the eigenvectors ψ_1, and ψ_2 is of no physical significance. This means we can extract a common phase factor from either of these equations or, to put it differently, we can assume one of the two coefficients in each equation to be real. Since γ in Eqs. 11.4 and 11.5 is real, this means that all four coefficients can be taken to be real. Thus we have

$$a_{11} = \gamma a_{21}, \qquad a_{22} = -\gamma a_{12} \qquad (11.9)$$

and, from Eq. 11.8,

$$a_{11}{}^2 + a_{12}{}^2 = 1 \qquad \text{and} \qquad a_{11}a_{21} + a_{12}a_{22} = 0 \qquad (11.10)$$

Solving these four equations we get the following values for the four coefficients:

$$a_{11} = -a_{22} = \frac{\gamma}{\sqrt{1+\gamma^2}}, \qquad a_{12} = a_{21} = \frac{1}{\sqrt{1+\gamma^2}} \qquad (11.11)$$

Hence

$$\psi_1 = \frac{1}{\sqrt{1+\gamma^2}}(\gamma\psi_T + \psi_S) \qquad \text{and} \qquad \psi_2 = \frac{1}{\sqrt{1+\gamma^2}}(\psi_T - \gamma\psi_S) \qquad (11.12)$$

A look at Eq. 11.4 shows us that

$$\lim_{x\to 0} \gamma = -\infty \qquad \text{and} \qquad \lim_{x\to\infty} \gamma = -1 \qquad (11.13)$$

In the absence of a magnetic field ($x = 0$), thus we have

$$\psi_1 = -\psi_T \qquad \text{and} \qquad \psi_2 = \psi_S \qquad (11.14)$$

as it should be. In a very strong magnetic field we get

$$(a) \quad \psi_1 = \frac{1}{\sqrt{2}}(-\psi_T + \psi_S) \qquad \text{and} \qquad (b) \quad \psi_2 = \frac{1}{\sqrt{2}}(\psi_T + \psi_S) \qquad (11.15)$$

Using the definitions of ψ_T and ψ_S, Eqs. 10.104 and 10.106 and Eqs. 11.15a and 11.15b become

$$(a) \quad \psi_1 = -\beta^+\alpha^- \qquad \text{and} \qquad (b) \quad \psi_2 = \alpha^+\beta^- \qquad (11.16)$$

This bears out the statement made in Chapter 7.4 that in a strong external field it is every spin for itself. A strong field decouples the two spins and makes m_I and m_j good quantum numbers. We started this paragraph with the intention of calculating the lifetimes of the states ψ_1 and ψ_2 as a function of the external magnetic field. Having followed the discussion to this point, the reader should now be able to carry out the final steps of this calculation.[4]

11.4   TWO- OR THREE-QUANTUM DECAY?

We have stated earlier without proof that singlet positronium decays into an even number of photons, usually two, whereas triplet positronium can

[4] See Problem 11.1.

decay only into an odd number of photons, usually three. This is due to the charge conjugation invariance of the electromagnetic interactions.

Charge conjugation means: Replace every particle in a system with its antiparticle.[5]

Charge conjugation invariance means: Nothing is changed by charge conjugation, i.e., a process that happens in the real world would happen in *exactly* the same way in an antiworld created by exchanging every particle with its antiparticle.

The electromagnetic interaction responsible for the annihilation of positronium and the strong (nuclear) interaction are charge conjugation invariant. The weak interaction that brings about nuclear β-decay, is not. To investigate the consequences of charge conjugation invariance for the annihilation of positronium we have to invoke another general principle: The wave function of a system of several identical fermions, i.e., particles of noninteger spin is antisymmetric under the exchange of particles. This means that the wave function of a system of several *identical* fermions (for example, several electrons) changes its sign if we exchange the coordinates of two particles:

$$\psi(\mathbf{r}_1, \mathbf{r}_2, \ldots, \mathbf{r}_i, \mathbf{r}_k, \ldots, \mathbf{r}_n) = -\psi(\mathbf{r}_1, r_2, \ldots, \mathbf{r}_k, \mathbf{r}_i, \ldots \mathbf{r}_n) \quad (11.17)$$

This theorem follows from field theoretical arguments and is a generalization of the Pauli principle. This latter fact we shall illustrate with an example. In Chapter 7.4, we had stated that the Pauli principle does not allow helium to have a triplet ground state. Let us now forget about this statement and instead insist that the ground-state eigenfunction be antisymmetric under particle exchange. We have not derived an explicit expression for the spatial part of the He eigenfunctions. We know, however, that in the ground state both electrons have the same quantum numbers ($n = 1, l = m = 0$) and, thus, the same eigenfunctions. Under an exchange of electron coordinates, therefore, the spatial part of the eigenfunction remains unchanged. Hence the antisymmetry must reside in the spin eigenfunction. Let α_1, β_1 and α_2, β_2 be the possible spin states of electron #1 and electron #2. Obviously there is only one normalized *non-zero* spin eigenfunction that we can form from these functions that is *antisymmetric*, under particle exchange:

$$\psi = \frac{1}{\sqrt{2}} (\alpha_1\beta_2 - \alpha_2\beta_1) \quad (11.18)$$

This is the singlet eigenfunction required by the Pauli principle. We can extend this antisymmetry requirement to the two *different* fermions forming

---

[5] This operation is sometimes more correctly referred to as particle-antiparticle conjugation, but the term charge conjugation is more widely used.

the positronium atom. We simply say: Positron and electron are two states of the *same* fermion, differing in one quantum number—*the charge.* If this is so, we have two identical particles *but* have to insist that the eigenfunctions are antisymmetric under exchange of the two "electrons": No. 1 and No. 2. In this new picture, let electron No. 1 have a positive charge and electron No. 2 a negative charge. An eigenfunction that is antisymmetric under charge exchange can be constructed as follows

$$\psi = \psi_1(\mathbf{r}, s, +)\psi_2(\mathbf{r}, s', -) - \psi_2(\mathbf{r}, s, +)\psi_1(\mathbf{r}, s', -) \qquad (11.19)$$

where the labels 1 and 2 refer to the "electrons" No. 1 and No. 2. Their spins and charges are respectively described by the symbols s, s' and $+$, $-$. Obviously Eq. 11.19 is antisymmetric under particle exchange (exchange of labels 1 and 2), as it should be, but what happens to it under charge exchange alone? We express charge conjugation with an operator S_c which changes the "charge quantum number" *plus* into *minus* and vice versa:[6]

$$S_c\psi = \psi_1(\mathbf{r}, s, -)\psi_2(\mathbf{r}, s', +) - \psi_2(\mathbf{r}, s, -)\psi_1(\mathbf{r}, s', -) \qquad (11.20)$$

We compare this with the influence of a "spin exchange operator," which exchanges only particle spins

$$S_p\psi = \psi_1(\mathbf{r}, s', +)\psi_2(\mathbf{r}, s, -) - \psi_2(\mathbf{r}, s', +)\psi_1(\mathbf{r}, s, -) \qquad (11.21)$$

By comparison we find

$$S_c\psi = -S_p\psi \qquad (11.22)$$

Hence in order to see what the charge conjugation operator does to the antisymmetrized wave function, we need only to see what spin exchange does. This depends, of course, on the spin eigenfunction:

$$S_p\alpha^+\alpha^- = \alpha^+\alpha^- \qquad (11.23)$$

$$S_p\psi_T = S_p\frac{1}{\sqrt{2}}(\alpha^+\beta^- + \alpha^-\beta^+) = \frac{1}{\sqrt{2}}(\beta^+\alpha^- + \beta^-\alpha^+) = \psi_T \qquad (11.24)$$

$$S_p\beta^+\beta^- = \beta^+\beta^- \qquad (11.25)$$

but

$$S_p\psi_S = S_p\frac{1}{\sqrt{2}}(\alpha^+\beta^- - \alpha^-\beta^+) = \frac{1}{\sqrt{2}}(\beta^+\alpha^- - \beta^-\alpha^+) = -\psi_S \qquad (11.26)$$

Hence

$$S_c\varphi_T = -\varphi_T \qquad (11.27)$$

and

$$S_c\varphi_S = \varphi_S \qquad (11.28)$$

[6] We assume that the positronium atom is in an s state so that the radial part of the eigenfunction is not dependent on the labeling of the particles.

where φ_T and φ_S are the complete eigenfunctions of the triplet and singlet ground states of positronium. φ_T and φ_S describe the system *before* annihilation; after annihilation it is described by

$$\varphi = \varphi_1 \cdot \varphi_2 \cdot \varphi_3 \cdots \tag{11.29}$$

the product of the wave functions φ_1, φ_2, etc., of the photons emitted.

What does a photon do under charge conjugation? A photon is its own antiparticle, *but* if we change all positive charges in the world into negative ones and vice versa, the electric and magnetic vectors of all photons change sign. This leaves the frequency and direction of the photon unchanged, but changes its phase by 180° and thereby changes the sign of its wave function. Thus φ_k goes to $-\varphi_k$ and φ remains unchanged if we have an even number of photons, but changes sign if the number of photons is odd.

Therefore, equations 11.27 to 11.29, taken together, say: If the electromagnetic interaction causing the annihilation is invariant under charge conjugation (and it is), then the triplet state positronium must decay into an odd number of photons and the singlet state positronium must decay into an even number of photons.

PROBLEMS

11.1 (a) Derive an expression for the lifetime of the two positronium hyperfine states with $m_f = 0$ as a function of the magnetic field.
(b) Discuss the quantitative results in comparison with the qualitative statements made in Chapter 11.3.

11.2 You are given the task of verifying the prediction made by quantum electrodynamics that the lifetime of the state ψ_s is 1.25×10^{-10} sec. Lifetimes of this order are very hard to measure accurately. How would you go about your assignment?

11.3 What is the energy difference between the positronium hyperfine states $(f, m_f) = (1, 1)$ and $(f, m_f) = (1, 0)$ in a magnetic field of 10,000 gauss?

11.4 The lifetime of the triplet positronium states $(f, m_f) = (1, 0)$, $(1, \pm 1)$ in solids (for instance plastic insulators) is reduced to $\approx 2 \times 10^{-9}$ sec because of the annihilation of the positron with bound electrons in the material and/or collision-induced spin flips to the singlet state. Calculate the strength of the magnetic field needed to reduce the lifetime of the $(1, 0)$ triplet state to 10^{-9} sec. (Assume that the reduction of the lifetime in the magnetic field takes place only as a result of state mixing.)

11.5 It is technically difficult to measure directly the relatively large energy difference between the singlet and the triplet state of positronium. In several experiments performed to date, experimenters have instead measured the energy difference between the states $(f, m_f) = (1, 0)$ and $(f, m_f) = (1, 1)$

in a magnetic field and have calculated the zero field splitting from the Breit-Rabi formula. In doing this, they face a dilemma: In a weak magnetic field the lifetime of the state $(1, 0)$ is relatively large, but the energy difference ΔE (between the states $(1, 0)$ and $(1, 1)$) is small, giving rise to large relative errors. In a strong field ΔE is large, but the lifetime of the $(1, 0)$ state is reduced, giving rise to a large uncertainty in ΔE because of the uncertainty principle. If our goal is to determine the zero field energy difference in the customary way, what magnetic field would allow the highest accuracy? Is this choice of field critical?

11.6 Calculate the energy difference between the positronium hyperfine states $(f, m_f) = (1, \pm 1)$ and $(f, m_f) = (1, 0)$ in an external magnetic field of 10 k gauss. Calculate the energy difference between the states $(f, m_f) = (1, \pm 1)$ and $(f, m_f) = (0, 0)$ in the same field.

11.7 You want to measure the energy difference between the positronium hyperfine states $(f, m_f) = (1, +1)$ and $(f, m_f) = (1, 0)$ with a relative accuracy of ± 1.5 percent. The magnet available to you has an inhomogeneity of ± 1 percent independent of the magnetic field it produces. What is the minimum value of the magnetic field needed to achieve the desired relative accuracy?

Solution of 11.7. The energy difference between the states $(f, m_f) = (1, \pm 1)$ and $(f, m_f) = (1, 0)$ is (according to Eq. 10.131)

$$\epsilon = -\frac{\Delta E}{4} + \frac{\Delta E}{2}\sqrt{1 + x^2} - \frac{\Delta E}{4} = \frac{\Delta E}{2}(\sqrt{1 + x^2} - 1)$$

The uncertainty in ϵ because of the uncertainty in B (or x) is

$$\Delta\epsilon = \frac{\partial\epsilon}{\partial x}\Delta x = \frac{\Delta E}{2}\frac{x}{\sqrt{1 + x^2}}\Delta x$$

Thus the relative uncertainty in ϵ expressed in terms of the relative uncertainty of x is

$$\frac{\Delta\epsilon}{\epsilon} = \frac{\Delta E}{2}\frac{x}{\sqrt{1 + x^2}}\frac{\Delta x}{\frac{\Delta E}{2}(\sqrt{1 + x^2} - 1)} \cdot \frac{x}{x}$$

$$= \frac{x^2}{1 + x^2 - \sqrt{1 + x^2}} \cdot \left(\frac{\Delta x}{x}\right)$$

Letting $\Delta x/x = 0.01$ and $\Delta\epsilon/\epsilon = 0.015$ we obtain

$$\frac{3}{2} = \frac{x^2}{1 + x^2 - \sqrt{1 + x^2}}$$

or

$$x = \sqrt{3}$$

Using the definition of x and the values of g and μ_0, the reader will obtain the value of B.

12

TIME DEPENDENT PERTURBATION THEORY

12.1 THE FORMALISM

Frequently one has a perturbation that changes with time, and the perturbation theory we have developed can be modified to deal with such a situation. Let us assume that a system is described by a Hamiltonian

$$H = H^{(0)} + H^{(1)}(t) \qquad (12.1)$$

where $H^{(0)}$ is the unperturbed Hamiltonian and $H^{(1)}(t)$ is a *small* time-dependent perturbation. The unperturbed Hamiltonian has eigenfunctions u_k

$$H^{(0)}u_k = E_k u_k \qquad (12.2)$$

which are the solutions of the time-independent Schrödinger equation. Since H is time dependent, its eigenfunctions must be time dependent, and we have to solve the *time-dependent Schrödinger equation:*

$$H\psi(t) = i\hbar \frac{\partial \psi}{\partial t} \qquad (12.3)$$

We can expand ψ in terms of the solutions of the unperturbed Schrödinger equation but, if we do this, the expansion coefficients must, of course, be functions of time:

$$\psi(t) = \sum_k a_k(t) u_k e^{-iE_k t/\hbar} \qquad (12.4)$$

We substitute Eqs. 12.4 and 12.1 into Eq. 12.3:

$$(H^{(0)} + H^{(1)})\psi(t) = \sum_k (H^{(0)} + H^{(1)}) a_k(t) u_k e^{-iE_k t/\hbar}$$

$$= i\hbar \left[\sum_k \dot{a}_k(t) u_k e^{-iE_k t/\hbar} - \sum_k a_k(t) u_k \frac{iE_k}{\hbar} e^{-iE_k t/\hbar} \right] \qquad (12.5)$$

Because of Eq. 12.2 this becomes

$$\sum_k \cancel{E_k a_k(t) u_k e^{-iE_k t/\hbar}} + \sum_k H^{(1)}(t) a_k(t) u_k e^{-iE_k t/\hbar}$$
$$= i\hbar \sum_k \dot{a}_k(t) u_k e^{-iE_k t/\hbar} + \cancel{\sum_k a_k(t) E_k u_k e^{-iE_k t/\hbar}} \quad (12.6)$$

We multiply Eq. 12.6 from the left with $u_n{}^*$ and integrate over the volume:

$$\sum_k a_k(t) e^{-iE_k t/\hbar}(u_n, H^{(1)}(t) u_k) = i\hbar \sum_k \dot{a}_k(t) e^{-iE_k t/\hbar}\delta_{nk} = i\hbar \dot{a}_n(t,e^{-iE_n t/\hbar} \quad (12.7)$$

On the right-hand side of Eq. 12.7 we have made use of the orthonormality of the u_k. Introducing

$$\frac{E_n - E_k}{\hbar} = \omega_{nk} \quad (12.8)$$

we can rewrite Eq. 12.7:

$$\dot{a}_n(t) = \frac{1}{i\hbar} \sum_k a_k(t) e^{i\omega_{nk}t} H_{nk}^{(1)} \quad (12.9)$$

In the most general case this is a system of infinitely many coupled linear differential equations, but we can solve it approximatively. To this end we assume that prior to the time $t = 0$, the system is in one of the unperturbed states u, and we call this particular state u_{initial} or, for short, u_i. At $t = 0$, the perturbation is turned on. If the perturbation is small (as we shall assume) the coefficients a_k in Eq. 12.9 will be small. This means $\dot{a}_n$ is small or, in other words, the coefficients a_n change only slowly with time. We are thus justified in assuming that, for some time after the perturbation has been turned on, the a_k on the right-hand side of Eq. 12.9 will remain close to their initial values $a_k(0)$. To determine the $a_k(0)$, we use Eq. 12.4, letting $t = 0$:

$$\psi(0) = u_i = \sum_n a_n(0) u_n \quad (12.10)$$

we multiply from the left with $u_k{}^*$ and integrate:

$$\delta_{ki} = a_k(0) \quad (12.11)$$

Now we can solve Eq. 12.9:

$$a_n(t) = \frac{1}{i\hbar} \int_{-\infty}^{\infty} e^{i\omega_{ni}t'} H_{ni}^{(1)} \, dt' \quad (12.12)$$

The reader, who is bothered by the fact that we assume the a_k to be constant in order to determine their time dependence, may convince himself that a more formal approach, letting $H^{(1)} = \epsilon H^{(1)\prime}$ and

$$a_k(t) = a_k^{(0)} + \epsilon a_k^{(1)} + \epsilon^2 a_k^{(2)} + \cdots \quad (12.13)$$

where $\epsilon \ll 1$, yields the same result if one neglects all terms of second or higher order in ϵ.

In the following paragraphs we shall apply this formalism to some simple problems. The first two are perturbation problems, all right, but we shall be able to diagonalize the perturbation Hamiltonian exactly (i.e., we shall not need to make approximations in solving Eq. 12.9.)

12.2 SPIN PRECESSION

If the perturbation mixes only a finite number of states, for example two, and if the time dependence of the Hamiltonian is simple, Eq. 12.9 can be solved exactly. As an example we treat the case of a free electron at rest. In the α-state the electron spin is assumed to point in the z-direction. At the time $t = 0$, a magnetic field $\mathbf{B}$ in the x-direction is switched on and stays on thereafter. The unperturbed states in this case are obviously

$$\alpha = \begin{pmatrix} 1 \\ 0 \end{pmatrix} \quad \text{and} \quad \beta = \begin{pmatrix} 0 \\ 1 \end{pmatrix} \tag{12.14}$$

and the eigenvector of the initial state is by assumption

$$\psi_i = \alpha \tag{12.15}$$

The unperturbed Hamiltonian is zero since we had assumed the particle to be at rest; the perturbation Hamiltonian is

$$\mathbf{H} = B_x g \cdot \mu_o \frac{\sigma_x}{2} = \mu_o B_x \sigma_x \tag{12.16}$$

(letting $g = 2$). Thus Eq. 12.9 becomes

$$\dot{a}_1(t) = \frac{\mu_o B_x}{i\hbar} [a_1(t)(\alpha_1, \sigma_x \alpha) + a_2(t)e^{i\omega_{12}t}(\alpha, \sigma_x \beta)] \tag{12.17}$$

and

$$\dot{a}_2(t) = \frac{\mu_o B_x}{i\hbar} [a_1(t)e^{i\omega_{21}t}(\beta, \sigma_x \alpha) + a_2(t)(\beta, \sigma_x \beta)] \tag{12.18}$$

The unperturbed states have zero energy, hence

$$\omega_{12} = \omega_{21} = 0$$

Now, according to Eq. 10.95,

$$\sigma_x \alpha = \beta \quad \text{and} \quad \sigma_x \beta = \alpha$$

Hence

$$\dot{a}_1(t) = \frac{\mu_o B_x}{i\hbar} a_2(t) \tag{12.19}$$

and

$$\dot{a}_2(t) = \frac{\mu_o B_x}{i\hbar} a_1(t) \tag{12.20}$$

We differentiate Eq. 12.19 and substitute it into Eq. 12.20,

$$\frac{i\hbar}{\mu_o B_x} \ddot{a}_1(t) = \frac{\mu_o B_x}{i\hbar} a_1(t) \tag{12.21}$$

or

$$\ddot{a}_1(t) = -\left(\frac{\mu_o B_x}{\hbar}\right)^2 a_1(t) \tag{12.22}$$

This has the solution:

$$a_1(t) = ae^{i\mu_o B_x t/\hbar} + be^{-i\mu_o B_x t/\hbar} \tag{12.23}$$

Differentiation of Eq. 12.23 and substitution into Eq. 12.19 yields

$$a_2(t) = -(ae^{i\mu_o B_x t/\hbar} - b \cdot e^{-i\mu_o B_x t/\hbar}) \tag{12.24}$$

From

$$\psi(0) = \alpha = a_1(0)\alpha + a_2(0)\beta \tag{12.25}$$

follows

$$a = b = \tfrac{1}{2} \tag{12.26}$$

hence

$$a_1(t) = \cos\left(\frac{\mu_o B_x t}{\hbar}\right) \tag{12.27}$$

and

$$a_2(t) = -i \sin\left(\frac{\mu_o B_x t}{\hbar}\right) \tag{12.28}$$

Thus the spin eigenvector of an electron in a magnetic field becomes

$$\psi(t) = \cos\left(\frac{\mu_o B_x t}{\hbar}\right)\alpha - i \sin\left(\frac{\mu_o B_x t}{\hbar}\right)\beta \tag{12.29}$$

i.e., the perturbation mixes the two states α and β and the degree to which each state contributes to the total state vector changes with time. According to Eq. 12.29, the spin stays in the y-z plane but precesses around the direction of the magnetic field with the Larmor frequency:

$$\omega = \frac{\mu_o B_x}{\hbar} \tag{12.30}$$

This result could also have been obtained by using a semiclassical approach. Our reason for giving this trivial problem the full treatment is to demonstrate the formalism on a simple situation before we graduate to something more complicated. Equation 12.29 also gives us another look at the mixing of states.

12.3 THE MIXING OF STATES

We compare Eq. 12.29 with the result given by stationary perturbation theory.

The energy eigenvalues are given according to Eq. 9.36 by

$$\begin{vmatrix} -E_k & \mu_o B_x \\ \mu_o B_x & -E_k \end{vmatrix} = 0 \quad \text{or} \quad E_k = \pm \mu_o B_x \quad (12.31)$$

Using these values of E_k in Eq. 9.35, we get

(a) $\mu_o B a_{21}{}^* = a_{11}{}^* \mu_o B \quad \text{or} \quad a_{21}{}^* = a_{11}{}^*$

(b) $\mu_o B a_{22}{}^* = -a_{12} \mu_o B \quad \text{or} \quad a_{22}{}^* = -a_{12}{}^*$ (12.32)

From the unitarity of

$$\mathbf{A} = \begin{pmatrix} a_{11} & a_{12} \\ a_{11} & -a_{12} \end{pmatrix} \quad (12.33)$$

it follows that

$$|a_{11}| = |a_{12}| = \frac{1}{\sqrt{2}}$$

Hence

$$\psi = \frac{1}{\sqrt{2}} (\alpha \pm \beta) \quad \text{with eigenvalues } \pm \mu_o B_x \quad (12.34)$$

Comparing Eqs. 12.29 and 12.34, we recognize another possible interpretation of the mixing of states. Under the influence of the perturbation the system oscillates back and forth between the various unperturbed states mixed together by the perturbation. The frequency of this oscillation is determined by the strength of the perturbation.[1] *Stationary perturbation theory* tells us the *time average* of the probability of finding the system in one state or another.

To illustrate the oscillation of a system between two states under the influence of a perturbation, we consider the decay of muons in a magnetic field. Muons have spin $\frac{1}{2}$ and thus behave in a magnetic field exactly as described in Chapter 12.2. As a result of the nonconservation of parity in the $\pi \to \mu + \nu$ decay producing the muon, it is easy to obtain beams of muons that are almost completely polarized along the direction of flight.[2]

[1] The question whether all mixed states can be interpreted in this way is somewhat academic since a stationary perturbation usually exists all along, and we do not know the unperturbed initial state.

[2] To show that the production of polarized muons in the decay of the meson requires nonconservation of parity in the decay process is not difficult. There are several very readable accounts of the connection between parity nonconservation and asymmetries in the angular distribution of the decay products. See, for instance, E. Wigner. *Scientific American* (December 1965), p. 28.

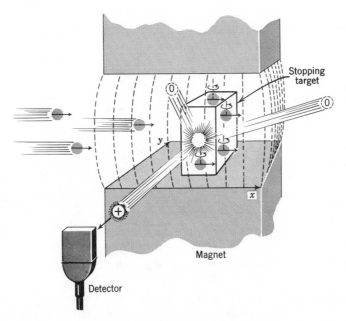

Fig. 12.1 Spin precession. The original axis of quantization was the direction of flight. With respect to this axis the muons were polarized. With respect to the direction defined by the magnetic field the muons are in a mixed state.

Let us assume now that a short burst of *polarized* muons is shot into a target and stops there (Figure 12.1).

A muon slows down and comes to rest (or rather thermal equilibrium) in a very short time ($\approx 10^{-10}$ sec). Its polarization is not disturbed in the process and, if the target is made of, say, graphite, there are no interactions to disturb the polarization afterward. Thus if the spins of all the muons point originally in the direction of the beam, they will continue to do so until the particles decay. This situation changes if the target is in a magnetic field in the z-direction. The stopped muons will, starting at $t = 0$, precess around the z-direction, and the spins which originally all pointed in the x-direction will point in succession in the $+y$, $-x$, $-y$, etc., direction.

Positive muons decay with a lifetime of 2.212 μsec into a positron and two neutrinos, and they have the fortunate habit of emitting the positron predominantly in the direction opposite to where their spin pointed at the moment of decay. This gives us a means of measuring the actual spin direction of the muons as a function of time. A detector (see Figure 12.1) registers the number of the decay positrons emitted at an angle between θ and $\theta + \Delta\theta$. If we plot the countrate of dN/dt as a function of the time τ elapsed

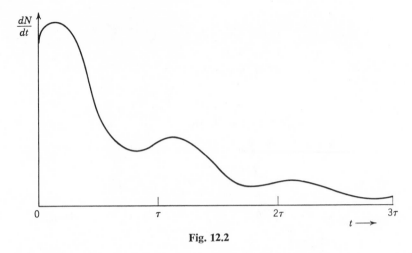

Fig. 12.2

since the arrival of the short burst of polarized muons, we find the time dependence shown in Figure 12.2.

Initially most of the positrons are emitted in the x-direction; after a $\frac{1}{4}$ turn of the spin, most decay positrons are emitted in the direction of the detector, etc. The decreasing amplitude of the curve, of course, results from the fact that the total number of muons decreases continuously with the lifetime of 2.212 μsec.

12.4 HARMONIC PERTURBATIONS

Very frequently one has to deal with perturbations caused by an oscillating electric or magnetic field. Using the formalism developed in Chapter 12.1, we can treat such problems in a general manner, but we shall restrict ourselves here to an example which is of special interest to experimental physicists.

A free electron (or other spin $\frac{1}{2}$ particle) whose spin is pointing in the z-direction is at rest in a constant magnetic field $\boldsymbol{B} = \boldsymbol{k}B = B_z$, also pointing in the z-direction. At the time $t = 0$, a small rotating magnetic field

$$\boldsymbol{B} = B \begin{pmatrix} \cos{(\omega_0 t)} \\ \sin{(\omega_0 t)} \\ 0 \end{pmatrix} \tag{12.35}$$

in the x-y plane is turned on. The Hamiltonian of the system is obviously[3]

$$\boldsymbol{H} = \boldsymbol{H}^{(0)} + \boldsymbol{H}^{(1)} = \frac{\mu_o g}{2} B_z \boldsymbol{\sigma}_z + \frac{\mu_o g}{2} B(\cos{\omega_o t}, \sin{\omega_o t}, 0) \begin{pmatrix} \sigma_x \\ \sigma_y \\ \sigma_z \end{pmatrix} \tag{12.36}$$

[3] Concerning the notation used here (i.e., row and column vectors), see Problem A.12 of the Appendix.

Letting $g = 2$, we obtain

$$\mathbf{H} = \mu_o B_z \sigma_z + \mu_o B[\cos(\omega_o t)\sigma_x + \sin(\omega_o t)\sigma_y] \tag{12.37}$$

The unperturbed state of the system, prior to $t = 0$, is described by the eigenvectors

$$\alpha \quad \text{and} \quad \beta$$

with eigenvalues given by

$$\mathbf{H}^{(0)}\alpha = \mu_o B_z \alpha \quad \text{and} \quad \mathbf{H}^{(0)}\beta = -\mu_o B_z \beta \tag{12.38}$$

The eigenvector of the perturbed state is

$$\psi(t) = a_1(t)\alpha + a_2(t)\beta \tag{12.39}$$

We had assumed that the electron was initially in the state α, hence

$$\psi(0) = a_1(0)\alpha + a_2(0)\beta = \alpha \tag{12.40}$$

or

$$a_1(0) = 1; \quad a_2(0) = 0 \tag{12.41}$$

Applying $\mathbf{H}^{(1)}$ to α and β, we get according to Eq. 12.37

$$\mathbf{H}^{(1)}\alpha = \mu_o B[\cos(\omega_o t)\beta + i\sin(\omega_o t)\beta] = \mu_o B e^{i\omega_o t}\beta \tag{12.42}$$

similarly

$$\mathbf{H}^{(1)}\beta = \mu_o B e^{-i\omega_o t}\alpha \tag{12.43}$$

Hence it follows that

$$H_{11}^{(1)} = H_{22}^{(1)} = 0 \tag{12.44}$$

$$H_{12}^{(1)} = H_{\alpha,\beta}^{(1)} = B\mu_o e^{-i\omega_o t} \tag{12.45}$$

and

$$H_{21}^{(1)} = H_{\beta,\alpha}^{(1)} = B\mu_o e^{i\omega_o t} \tag{12.46}$$

Using Eqs. 12.44 to 12.46 in Eq. 12.9, we obtain

$$\dot{a}_1(t) = \frac{1}{i\hbar} a_2(t)\mu_o B e^{i\omega_{12}t} e^{-i\omega_o t} \tag{12.47}$$

and

$$\dot{a}_2(t) = \frac{1}{i\hbar} a_1(t)\mu_o B e^{i\omega_{21}t} e^{i\omega_o t} \tag{12.48}$$

Using

$$\frac{E_1 - E_2}{\hbar} = \omega_{12} = -\omega_{21} = \omega \tag{12.49}$$

this becomes

$$\dot{a}_1(t) = \frac{\mu_o B}{i\hbar} a_2(t)e^{i(\omega-\omega_o)t} \tag{12.50}$$

and

$$\dot{a}_2(t) = \frac{\mu_o B}{i\hbar} a_1(t) e^{-i(\omega - \omega_o)t} \tag{12.51}$$

Differentiation of Eq. 12.51 and substitution into Eq. 12.50 yields

$$\frac{i\hbar}{\mu_o B} \{\ddot{a}_2(t) + i(\omega - \omega_o)\dot{a}_2(t)\} e^{i(\omega - \omega_o)t} = \frac{\mu_o B}{i\hbar} a_2(t) e^{i(\omega - \omega_o t)} \tag{12.52}$$

This and the differentiation of Eq. 12.50 and its substitution into Eq. 12.51 yields two differential equations:

$$\ddot{a}_2(t) + i(\omega - \omega_o)\dot{a}_2(t) + \left(\frac{\mu_o B}{\hbar}\right)^2 a_2(t) = 0 \tag{12.53}$$

and

$$\ddot{a}_1(t) - i(\omega - \omega_o)\dot{a}_1(t) + \left(\frac{\mu_o B}{\hbar}\right)^2 a_1(t) = 0 \tag{12.54}$$

We substitute

$$a_2(t) = e^{i\gamma t} \tag{12.55}$$

into Eq. 12.53 and obtain

$$-\gamma^2 - \gamma(\omega - \omega_o) + \left(\frac{\mu_o B}{\hbar}\right)^2 = 0 \tag{12.56}$$

This has the solution:

$$\gamma = \frac{\omega - \omega_o}{2} \pm \sqrt{\left(\frac{\omega - \omega_o}{2}\right)^2 + \left(\frac{\mu_o B}{\hbar}\right)^2} \equiv \frac{\Delta\omega}{2} \pm \lambda \tag{12.57}$$

Thus, the general solution of Eq. 12.53 is

$$a_2(t) = e^{i\Delta\omega t/2}(A \cdot e^{i\lambda t} + B e^{-i\lambda t}) \tag{12.58}$$

From the initial condition, $a_2(0) = 0$ follows $B = -A$. We differentiate Eq. 12.58 and let $t = 0$:

$$\dot{a}_2(0) = 2Ai\lambda \tag{12.59}$$

Equating this with Eq. 12.48, yields

$$A = -\frac{\mu_o B}{2\lambda\hbar} \tag{12.60}$$

Hence it follows that

$$a_2(t) = \frac{\mu_o B}{2\lambda\hbar} e^{i(\Delta\omega t)/2}(e^{-i\lambda t} - e^{i\lambda t}) = -i\frac{\mu_o B}{\lambda\hbar} \sin(\lambda t) \cdot e^{i(\Delta\omega t)/2} \tag{12.61}$$

Thus the probability of finding the electron in the state β at the time t is given by

$$|a_2(t)|^2 = \left(\frac{\mu_o B}{\lambda \hbar}\right)^2 \sin^2(\lambda t) \qquad (12.62)$$

Now, if at $t = 0$, the probability of finding the electron in the state β is zero, whereas a short time t later it is $|a_2(t)|^2$, then $|a_2(t)|^2$ must be the **transition probability** that during the time interval t the electron has gone from the state α to the state β. Using the value of λ as defined in Eq. 12.57, we obtain

$$|a_2(t)|^2 = \frac{(\mu_o B)^2}{\hbar^2\left[\frac{(\omega - \omega_o)^2}{4} + \left(\frac{\mu_o B}{\hbar}\right)^2\right]} \sin^2\left\{\sqrt{\frac{(\omega - \omega_o)^2}{4} + \left(\frac{\mu_o B}{\hbar}\right)^2}\, t\right\} \qquad (12.63)$$

We investigate this expression. The probability that the electron goes from the state α to the state β under the influence of a harmonic perturbation of frequency ω_o varies with time, and the amplitude of this variation is largest if

$$\omega = \omega_o = \frac{2\mu_o B_z}{\hbar} \qquad (12.64)$$

If we want the transition probability to equal unity, we also have to maximize the $\sin^2$ term in Eq. 12.63 by letting

$$\lambda t = \sqrt{\frac{(\omega - \omega_o)^2}{4} + \left(\frac{\mu_o B}{\hbar}\right)^2}\, t = \frac{\pi}{2} \qquad (12.65)$$

This result is often—colloquially—expressed as follows: The application of a magnetic field rotating with the *resonance frequency* $\omega_0 = \omega$ for a time $t = \pi/2\lambda$ *"flips the spin"* from the α to the β state. If we had left the perturbation on twice as long, $t = \pi/\lambda$, $|a_2|^2$ would again be zero, meaning that the spin would have been flipped back to the α state. The resonance character of Eq. 12.63 enables us to measure magnetic moments with great precision. The magnetic field B and/or the time t are usually adjusted in such a way that *at resonance*, i.e., when $\omega = \omega_0$, the transition probability is equal to one. We assume this to be the case and plot

$$\left|a_2\left(\frac{\pi\hbar}{2\mu_o B}\right)\right|^2 = \frac{1}{\frac{\hbar^2(\omega - \omega_o)^2}{4\mu_o{}^2 B^2} + 1} \sin^2\left\{\sqrt{1 + \frac{\hbar^2(\omega - \omega_o)^2}{4\mu_o{}^2 B^2}}\, \frac{\pi}{2}\right\} \qquad (12.66)$$

as a function of (see Figure 12.3):

$$\frac{(\omega - \omega_o)\hbar}{2\mu_o B}$$

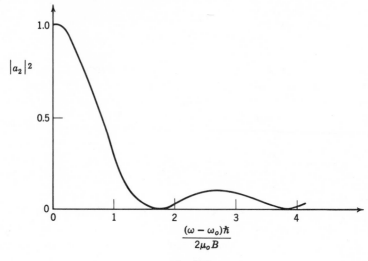

Fig. 12.3

We can, thus, from a measurement of ω_o and B_z determine the magnetic moment μ_o of a particle. If we increase B (i.e., increases the power of the rotating *r.f.* magnetic field), the curve in Figure 12.3 is broadened.

A larger *r.f.* magnetic field flips the spin in a shorter time; this so-called "power broadening" of the spectral line is, therefore, a manifestation of the uncertainty principle. Obviously the experiment can be performed in two completely equivalent ways: (a) by changing the *frequency* ω_o of the *r.f. magnetic field*, or (b) by changing the frequency ω through a change of the *field strength* B_z of the *static magnetic field*.

It is interesting to note that for a given *r.f.* magnetic field the transition probability from the state α to the state β is just as big as the one from the state β to the state α. This is generally true in quantum mechanics: A process that has a high transition probability of going in one direction also has a high probability of going in the opposite direction.[4]

Returning to Eq. 12.42, we notice that if we had written $-\sin \omega_o t$ instead of $\sin \omega_o t$, thus changing the direction of rotation of the magnetic field, we would have had to replace $\omega - \omega_o$ with $\omega + \omega_o$ in Eq. 12.57 and would not have obtained a resonance.

It should be noted that it is not necessary to use a rotating *r.f.* magnetic field in *magnetic resonance* experiments of this type. A linearly polarized *r.f.*

[4] See Problem 8.3.

magnetic field in the x-direction can be written as

$$\mathbf{B} = 2B\begin{pmatrix} \cos \omega_o t \\ 0 \\ 0 \end{pmatrix} = B\begin{pmatrix} \cos \omega_b t \\ \sin \omega_o t \\ 0 \end{pmatrix} + B\begin{pmatrix} \cos \omega_o t \\ -\sin \omega_o t \\ 0 \end{pmatrix} \quad (12.67)$$

(i.e., as a superposition of two fields—one rotating clockwise, the other rotating counterclockwise). As we have seen, the field that rotates clockwise makes a negligible nonresonant contribution. Thus the experiment can be performed with a linearly polarized magnetic field if we make the obvious adjustment of the amplitude.

Resonance experiments of this type have played a crucial role in the development of some branches of physics, and we shall now discuss a few typical (and important) ones.

12.5 MAGNETIC RESONANCE EXPERIMENTS

Molecular Beam Experiments

Stern and **Gerlach**[5] demonstrated the possibility of separating in an atomic beam the two spin states of silver atoms in the ground state. The accuracy of this method of measuring the magnetic moment of atoms and molecules was improved by **I. I. Rabi** by several orders of magnitude. A molecular beam is first split according to the components of the angular momentum of its molecules (atoms) in an inhomogeneous field A of the Stern-Gerlach type (see Figure 12.4). One component, respresenting, for example, a certain spin state, is then passed into a homogeneous field B. Superimposed on this homogeneous field is an r.f. magnetic field of variable frequency ω_o. From the field B the molecules pass into another inhomogeneous field C that deflects them onto a detector D. If the r.f. field is tuned to the energy difference (frequency) between the two spin states, the angular momentum of the molecules will change upon passage through the field B. In this case, the magnet C will no longer deflect them onto the detector D. The signal from the detector D will thus vary as the r.f. frequency is changed. Obviously this technique—as well as its theoretical foundation as given in Chapter 12.4—can be extended to more complicated situations. The Rabi apparatus, for instance, can be used to measure energy differences between hyperfine states (in this case the Hamiltonian (Eq. 12.36) must include the $(I \cdot j)$ interaction.

Electron Resonance and Nuclear Magnetic Resonance

The nuclei in many chemical compounds and the electrons in certain free radicals are magnetically so loosely coupled to the rest of the molecule that

[5] See footnote on p. 134.

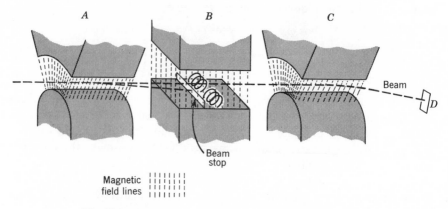

Fig. 12.4   A schematic of Rabi's molecular beams apparatus.

they can be considered free. In this case, the theory developed in Chapter 12.4 can be applied directly. In the experiment a liquid or solid specimen is placed in a strong homogeneous magnetic field superimposed on a weak r.f. field. The field strength of the d.c. field is then varied until the *resonance field*, as given by Eq. 12.64 is reached:

$$B_z = \frac{\hbar\omega}{2\mu} \tag{12.68}$$

(μ is here the magnetic moment of the particle under investigation). At this value of the d.c. field, spins are flipped and the r.f. circuit supplies the power:

$$P = N\hbar\omega \tag{12.69}$$

(N = number of spin flips/second) to the sample. At a certain setting of the d.c. field we can thus observe a change in the load of the r.f. circuit. Obviously this experiment requires a certain degree of polarization of the probe. If there were exactly the same number of particles in the energetically higher α-state as there are in the β-state, the r.f. circuit would gain as much energy from transitions from α to β as it would lose because of transitions from β to α (remember the transitions probability is the same for both processes). Fortunately we are saved by the fact that in thermal equilibrium there is a slightly higher population in the energetically lower state (and we mean slightly). At room temperature for fields in the kilo-gauss region the excess population in the β-state is only of the order of one part per million.

Since a sample easily contains more than 10^{20} molecules this is enough to give a measurable signal. Nuclear and electron magnetic resonance techniques (N.M.R. and E.M.R.) are used in very accurate magnetometers where the known magnetic moment of protons or electrons is used to measure d.c. magnetic fields by varying the frequency of the r.f. field until resonance occurs. Other applications include the measurement of small deviations from the free-particle resonance frequency (Eq. 12.64) that result from interaction with the other constituents of a molecule, helping in the determination of molecular structures.

Positronium

As mentioned in Chapter 11.2 the hyperfine structure of the positronium ground state has been measured. The reader should now be able to appreciate the elegance of the method as well as do all the pertinent calculations. Positronium is formed in a microwave cavity located in a d.c. magnetic field. A linearly polarized r.f. magnetic field in the cavity mixes the states $(f, m_f) = (1, \pm 1)$ and $(f, m_f) = (1, 0)$. Now, the state $(1, 0)$ is mixed by the d.c. field with the state $(0, 0)$. If the d.c. field is set to be off the resonance value,[6] all the atoms in the states $(1, 0)$ and $(0, 0)$ will decay into two quanta. As the resonance value is reached the two quantum decay channel is also opened to the atoms in the $(1, \pm 1)$ states, resulting in a change of the energy distribution of the annihilation γ-rays.

Since we are already well acquainted with the hyperfine structure of the positronium ground state, we shall carry out a detailed calculation of the transition probability for this particular example. The calculation will follow the procedure that we used for the determination of the transition probability between the two spin states of a free particle (see Chapter 12.4).

To keep things simple we shall assume that $x \ll 1$ (see Eq. 10.132), i.e., that the external magnetic field B_z is small. (The reader who settles for nothing short of perfection may obtain the complete solution by omitting this short cut.)

We write the Hamiltonian of the system in the following form

$$\mathbf{H} = \mathbf{H}^{(0)} + \mathbf{H}^{(1)} \tag{12.70}$$

where $\mathbf{H}^{(0)}$ is the time-independent part, including the interaction with the static magnetic field B_z. $\mathbf{H}^{(1)}(t)$ is the time-dependent perturbation caused by the rotating magnetic field:

$$\mathbf{B} = B \begin{pmatrix} \cos(\omega_o t) \\ \sin(\omega_o t) \\ 0 \end{pmatrix} \tag{12.71}$$

We want to obtain the matrix elements of the Hamiltonian $\mathbf{H}^{(1)}$ and thus must ask ourselves: What are the possible initial states of the system under the influence of the already present static field B_z? This question has already

---

[6] It is technically much easier to vary the d.c. field rather than vary the microwave frequency.

been answered in Chapter 11.3. The states $\alpha^+\alpha^-$ and $\beta^+\beta^-$ are eigenstates of $H^{(0)}$ and so are the states (Eq. 11.12)

$$\psi_1 = \frac{1}{\sqrt{1+\gamma^2}}(\gamma\psi_T + \psi_S) \quad \text{and} \quad \psi_2 = \frac{1}{\sqrt{1+\gamma^2}}(\psi_T - \gamma\psi_S) \quad (12.72)$$

making use of our assumption that $x \ll 1$ ψ_1 reduces to

$$\psi_1 = \frac{x}{2}\left(\psi_S - \frac{2}{x}\psi_T\right) = -\left(\psi_T - \frac{x}{2}\psi_S\right) \quad (12.73)$$

and Eq. (12.73) reduces to

$$\psi_2 = \frac{x}{2}\left(\psi_T + \frac{2}{x}\psi_S\right) = \psi_S + \frac{x}{2}\psi_T \quad (12.74)$$

The states $\alpha^+\alpha^-$ and $\beta^+\beta^-$ are degenerate so we could start with any linear combination. Which one should it be? To simplify matters we assume that all positrons are in the β state so that only the state $\beta^+\beta^-$ will be occupied.[7] Thus our initial states are

$$\beta^+\beta^-, \psi_1, \quad \text{and} \quad \psi_2$$

The Hamiltonian $\mathbf{H}^{(1)}$, written explicitly in analogy to Eq. 12.36, is

$$\mathbf{H}^{(1)} = \frac{\mu_o g B}{2}\left(\cos(\omega_o t), \sin(\omega_o t), 0\right)\begin{pmatrix} \sigma_x^- - \sigma_x^+ \\ \sigma_y^- - \sigma_y^+ \\ \sigma_z^- - \sigma_z^+ \end{pmatrix}$$

$$= \frac{\mu_o B g}{2}\left[(\sigma_x^- - \sigma_x^+)\cos(\omega_o t) + (\sigma_y^- - \sigma_y^+)\sin(\omega_o t)\right] \quad (12.75)$$

Fortunately we do not have to calculate all the matrix elements H_{ik}, although this would not be overly difficult. We have seen in Chapter 12.4 that a spin-flip transition under the influence of a magnetic field,[8] (and that is what we are concerned with here) happens only if the frequency ω_o of the rotating magnetic field is very close to

$$\frac{\Delta E}{\hbar} = \omega$$

[7] This assumption is not completely unjustified since positrons from the β-decay of a radioactive nucleus are at least partly polarized. Our assumption does not affect the population in the states ψ_1 and ψ_2. True, the initial capture of the positron will only lead to a state $\beta^+\beta^-$ or $\beta^+\alpha^-$, but $\beta^+\alpha^-$ is not an eigenstate of the hyperfine interaction $\sigma^+ \cdot \sigma^-$, which immediately mixes it with the $\alpha^+\beta^-$ state to form the states ψ_S and ψ_T.

[8] The same statement can be made for all *resonance* transitions induced by oscillating or rotating electromagnetic fields. Other examples are the absorption of light by an atom and the absorption of γ-rays by a nucleus, both going to an excited state in the process.

where ΔE is the energy difference between the final and the initial state. Let us assume we have adjusted our magnetic field B_z so that $\hbar\omega_o$ equals the energy difference between the states ψ_1 and $\beta^+\beta^-$. In that case only the matrix elements $(\beta^+\beta^-, H^{(1)}\psi_1) = H^{(1)}_{\beta\beta,1}$ and $(\psi_1, H^{(1)}\beta^+\beta^-) = H^{(1)}_{1,\beta\beta}$ will contribute and they are, as the reader can easily (?) verify,[9]

$$\left(-\left(\psi_T - \frac{x}{2}\psi_S\right), H^{(1)}\beta^+\beta^-\right) = H^{(1)}_{1,\beta\beta} = -\frac{\mu_o gBx}{\sqrt{8}} e^{-i\omega_o t} \quad (12.76)$$

From the hermiticity of $H^{(1)}$ follows

$$H^{(1)}_{\beta\beta,1} = H^{(1)*}_{1,\beta\beta} = -\frac{\mu_o gB}{\sqrt{8}} e^{i\omega_o t} \quad (12.77)$$

Now,[10] using Eq. 12.9, we can calculate the coefficients $a_n(k)$.

$$\dot{a}_{\beta\beta}(t) = \frac{1}{i\hbar} a_1(t)e^{-i\omega t}H^{(1)}_{\beta\beta,1} = \frac{i\mu_o gBx a_1(t)}{\hbar\sqrt{8}} e^{i(\omega_o - \omega)t} \quad (12.78)$$

and

$$\dot{a}_1(t) = \frac{1}{i\hbar} a_{\beta\beta}(t)e^{i\omega t}H^{(1)}_{1,\beta\beta} = i\mu_o gBx a_{\beta\beta}(t)e^{-i(\omega_o - \omega)t} \quad (12.79)$$

This system of two coupled differential equations is, except for the values of the constants, the same as the one given by Eqs. 12.50 and 12.51. Therefore, we can obtain the solution immediately by substituting the proper values of the constants into Eq. 12.63. However, there is one hitch. Comparing Eqs. 12.78 and 12.79 with Eqs. 12.50 and 12.51, we find that $a_1(t)$ in Eq. 12.79 does not play the role of $a_2(t)$ in Eq. 12.51 but rather that of $a_2^*(t)$. Since we are only interested in the transition probability $|a_1|^2$ this need not bother us. Comparison of Eqs. 12.79 and 12.51 thus yields the substitution:

$$\frac{\mu_o B}{i\hbar} \to \frac{\mu_o gBx}{i\hbar\sqrt{8}} \quad (12.80)$$

In other words, in order to obtain the transition probability $|a_1|^2$ we must replace

$$B \text{ with } \frac{gBx}{\sqrt{8}}$$

in Eq. 12.63 obtaining

$$|a_1(t)|^2 = \frac{(\mu_o gBx)^2}{2\hbar^2\left[(\omega - \omega_o)^2 + \frac{1}{2}\left(\frac{gBx\mu_o}{\hbar}\right)^2\right]} \sin^2\left\{\sqrt{\frac{(\omega - \omega_o)^2}{4} + \frac{1}{8}\left(\frac{\mu_o gBx}{\hbar}\right)^2} \, t\right\}$$

---

[9] See Eqs. 10.95 to 10.97 and also Problem 12.4.
[10] See Problem 12.5.

So much for the transition probability. If, in an actual experiment, we want to observe a signal, it is necessary that there be a difference in the initial population of the two states being mixed by the time-dependent perturbation. This is where things get a little sticky. The population of the state $\beta^+\beta^-$ depends on the degree of polarization of the positrons, and the population of the state $\psi_1 = -(\psi_T - (x/2)\psi_S)$ depends on the magnetic field B_z. Also, the two states have a different lifetime and, to make things worse, the lifetime of one of the states ψ_1 depends on the magnetic field B_z. Nevertheless, the reader might be able to unravel even this part of the problem or, at least, follow the literature[11] on the subject.

We have treated this particular example in considerable detail to show how detailed information can be extracted from an experiment. Resonance experiments of the type described here have reached a very high level of sophistication. It is no longer enough for the experimental physicist to measure a line and then say, here it is. The small effects that result from nuclear forces or an interaction with the radiation field shift a resonance line sometimes only by a small fraction of the linewidth. Therefore, it is not unusual that experimenters determine the location of the center of a resonance line to one hundredth or even one thousandth of the linewidth. This means, of course, that the *lineshape*, i.e., the dependence of the transition probability on such things as magnetic field or frequency must be very well understood.

Muonium

A technique quite similar to that used for positronium has been used to measure the hyperfine structure of muonium. A resonance transition between the states $(f, m_f) = (1, 1)$ and $(1, 0)$ in a strong d.c. magnetic field $B = kB_z$ was induced by a rotating r.f. magnetic field. In the $(1, 1)$ state the muon is in an α state whereas in the $(1, 0)$ state it is in a mixed state. In this case, the resonance transition leads to an observable change in the angular distribution of the decay positrons.

PROBLEMS

12.1   The angular distribution of the positrons from muon decay is given by

$$P(\theta) = (1 - \tfrac{1}{3}\cos\theta) \qquad \text{(given without proof)}$$

$P(\theta)\,d\theta$ is the probability that the positron is emitted in the angular range between θ and $\theta + d\theta$. θ is the angle between the muon spin and the direction of flight of the positron.

[11] See for instance V. W. Hughes, S. Marder, and C. S. Wu. Phys. Rev. **106**, 934 (1957).

Derive an expression for the curve shown in Figure 12.2., assuming that the solid angle under which target and detector (Figure 12.1) see each other is very small.

12.2 (a) Derive an expression equivalent to Eq. 12.63 for the precession of a free proton.

(b) A free proton is located in a static magnetic field $\mathbf{B} = kB_z$ of 10,000 gauss. How strong a rotating r.f. resonant field must one apply in the x-y plane in order to "flip" the proton spin in 10^{-4} sec?

12.3 Muonium was discovered by measuring the precession frequency of the atom in a weak static magnetic field, i.e., the energy difference between the states $(f, m_f) = (1, 1)$ and $(1, -1)$. Compare this frequency with the precession frequency of a free muon in the same magnetic field. On which fact could the experimenters base their claim that muonium had, indeed, been formed? Does an experiment of this type give information about the size of the zero-field hyperfine structure splitting between the triplet and singlet state?

12.4 Derive Eqs. 12.76 and 12.77 explicitly.

12.5 Derive Eqs. 12.78 and 12.79.

12.6 Derive Eqs. 12.78 and 12.79, assuming that the initial state was not $\beta^+\beta^-$ but $\alpha^+\alpha^-$. Comment!

12.7 Derive all the elements of the matrix $\mathbf{H}^{(1)}$ in the representation defined by the states $\alpha^+\alpha^-$, ψ_1, $\beta^+\beta^-$, ψ_2.

Solution of 12.4. The Hamiltonian $\mathbf{H}^{(1)}$ is given in Eq. 12.75. We apply it to $\beta^+\beta^-$,

$$\frac{\mu_o g B}{2}\{(\sigma_x^- - \sigma_x^+)\cos(\omega_o t) + (\sigma_y^- - \sigma_y^+)\sin(\omega_o t)\}\beta^+\beta^- = \mathbf{H}^{(1)}\beta^+\beta^-$$

now

$$\sigma_x\beta = \alpha \quad \text{and} \quad \sigma_y\beta = -i\alpha \quad \text{(see Eqs 10.95 and 10.96)}$$

Hence

$$\mathbf{H}^{(1)}\beta^+\beta^- = \frac{\mu_o g B}{2}\{\cos(\omega_o t)(\beta^+\alpha^- - \alpha^+\beta^-) + i\sin(\omega_o t)(-\beta^+\alpha^- + \alpha^+\beta^-)\}$$

$$= \frac{\mu_o g B}{2}\{-\sqrt{2}\psi_S\cos(\omega_o t) + i\sqrt{2}\psi_S\sin(\omega_o t)\}$$

$$= \frac{-\mu_o g B}{\sqrt{2}}\psi_s\{\cos(\omega_o t) - i\sin(\omega_o t)\} = \frac{-\mu_o g B}{\sqrt{2}}\psi_s e^{-\omega_o t}$$

We multiply this with $\psi_1 = -\psi_T + x/2\,\psi_S$ and obtain, since ψ_T and ψ_S are orthogonal

$$\mathbf{H}^{(1)}_{1,\beta\beta} = \left(-\psi_T + \frac{x}{2}\psi_S\right)\left(\frac{-\mu_o g B}{\sqrt{2}}\psi_S e^{-i\omega_o t}\right) = \frac{-\mu_o g B}{\sqrt{8}}x e^{-i\omega_o t}$$

The reader may verify by explicit calculation that

$$\mathbf{H}^{(1)*}_{1,\beta\beta} = \mathbf{H}^{(1)}_{\beta\beta,1}$$

is indeed satisfied.

13

SCATTERING

13.1 THE FUNDAMENTALS

Thus far we have concerned ourselves mostly with particles held in potential wells, central potentials, or other entrapments. Now we shall investigate the fate that may befall a particle traveling through matter.

Let a parallel beam of monoenergetic particles be incident on a target containing scattering centers, i.e., atoms, nuclei, etc. (see Figure 13.1). Occasionally a particle will approach a target particle so closely that the two interact and the incident particle is *scattered*. The target is considered to be *thin* if the probability is small that a particle is scattered more than once upon

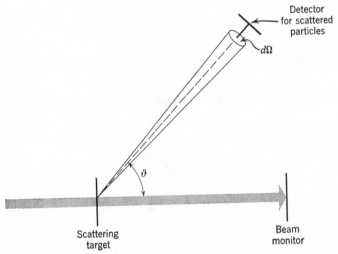

Fig. 13.1 The number of particles scattered into a detector is proportional to the solid angle $d\Omega$ covered by the detector, the number of incident particles as determined by the beam monitor, and the number of scattering centers in the target. (The fraction of particles scattered out of the beam is assumed to be small.) The proportionality factor is called the differential cross section.

traversing the target. In a typical scattering experiment one might measure the number of particles that are scattered by an angle ϑ into the element $d\Omega$ of the solid angle. This number will obviously be proportional to N' (the number of particles per cm² per sec incident upon the target), to n (the number of scattering centers in the target), and to the element $d\Omega$ of the solid angle.

Number of particles scattered per second into $d\Omega = \sigma(E, \vartheta)N'n\, d\Omega$ (13.1)

The proportionality factor $\sigma(E, \vartheta)$ can obviously depend on the energy E of the incident particles and on the scattering angle ϑ. If either the target or the beam particles are polarized, there can also be a dependence on the azimuth φ, a possibility that we shall henceforth disregard. The dimension of $\sigma(E, \vartheta)$ as defined by Eq. 13.1 is that of an area, and σ is called the differential cross section. The physical interpretation of σ is the following: if the incident particles were point particles and if every particle hitting a scattering center were to be scattered into $d\Omega$, each scattering center would have a cross section of σ [cm²].

Integration of $\sigma(E, \vartheta)$ over all angles yields the total cross section.

$$\sigma_o(E) = \int \sigma(E, \vartheta)\, d\Omega = \int \sigma(E, \vartheta) \sin \vartheta\, d\vartheta\, d\varphi \qquad (13.2)$$

It is obvious that it should be possible to calculate $\sigma(E, \vartheta)$ if the interaction between the target particles and the incident particles is known in detail. Similarly one should be able to derive salient features of the interaction from an analysis of scattering data.

Scattering experiments have been essential to the exploration of the forces acting between the various elementary particles. It should, however, be pointed out that scattering experiments are often difficult to interpret. During a collision the interaction is sampled over a wide range of distances and, maybe, relative velocities. Therefore, it is not surprising that a measured scattering distribution can usually be fitted with a variety of theoretical models. In the following paragraph we shall derive some of the features of potential scattering without ever committing ourselves to a specific potential for the scattering centers. Since scattering encompasses a wide variety of processes we narrow down the number of possibilities as follows.

1. The interaction between the particles can be described by a spherically symmetrical potential.

2. The scattering process is elastic, i.e., neither of the participating particles changes its rest mass and no particles are created or annihilated during the collision.

3. Target particles and incident particles are different. This means that we can always tell whether a particle emerging from the target is a scattered incident particle or a recoiling target particle.

13.2 THE SCHRÖDINGER EQUATION, PARTIAL WAVES

There can be no fundamental difference between the interaction of two bound particles and that of two free particles. We must, therefore, be able to start with the Schrödinger equation in the c.m. system (Eq. 5.20), written here (Eq. 13.3) for an arbitrary potential. The only difference in our procedure will be that we shall now seek solutions for unbound particles, i.e., those with $E > V$.

$$\frac{-\hbar^2}{2m} \nabla^2 u + Vu = Eu \tag{13.3}$$

$m = m_1 m_2 / m_1 + m_2$ is the reduced mass; m_1 we assume to be the mass of the incident particle and m_2 the mass of the target particle. To see just what the introduction of the reduced mass does for us in this case we calculate the energy in the center-of-mass system

$$E = \frac{m_1 v_1^2}{2} - \frac{(m_1 + m_2)v'^2}{2} \tag{13.4}$$

where v' is the velocity of the center of mass and v_1 the velocity of the incident particle in the laboratory system. The lab velocity v_2 of the target particle is, of course, zero. Hence

$$E = \frac{m_1 v_1^2}{2} - \frac{(m_1 + m_2)(m_1^2 v_1^2)}{2(m_1 + m_2)^2} = \frac{m_1 m_2 v_1^2}{2(m_1 + m_2)} = \frac{m v_1^2}{2} \tag{13.5}$$

This is the kinetic energy of a particle of mass m moving with the velocity v_1. Equations 13.3 thus describes the scattering of a particle of mass m and kinetic energy $m v_1^2/2$ by a fixed potential located at the center of mass.

To compare our findings with experimental results, we would have to transform the angles and cross sections to the laboratory system. The formulas for this routine task are derived in many texts.[1]

Even without solving Eq. 13.3 we can find out something about the general character of its solutions. In the immediate vicinity of the scattering center the action will be violent and its description difficult. At a considerable distance from the scattering target, however (where the experimentalist lies in wait for the scattered particles) things will be simpler, and a look at Figure 13.1 shows what the solutions will look like in this region.

There must be a plane wave that describes the unscattered particles and, superimposed upon it, a spherical wave that emanates from the target and

[1] For example, L. I. Schiff, *Quantum Mechanics*, second edition, McGraw-Hill, New York, 1955, p. 97 ff.

describes the scattered particles. We should thus be able to write the **asymptotic solution** of Eq. 13.3 as follows

$$u_\infty = \lim_{r \to \infty} u(\mathbf{r}_1 \vartheta) = A\left\{e^{ikz} + \frac{1}{r}f(\vartheta)e^{ikr}\right\} \tag{13.6}$$

where the first term describes the plane wave and the second term the spherical wave.

$$k = \frac{mv}{\hbar} \tag{13.7}$$

is the wave number of the (matter) wave. $f(\vartheta)$ is the amplitude of the spherical wave. It can, of course, still depend[2] on the angle ϑ and is usually called the **scattering amplitude**. Before we try to find $f(\vartheta)$ we derive an important theorem. The probability P of finding a particle in a volume V is given, according to p. 58, by

$$P = \int_V \psi^* \psi \, d\tau \tag{13.8}$$

We can calculate the rate of change of this probability

$$\frac{dP}{dt} = \int_V \left(\psi^* \frac{\partial \psi}{\partial t} + \psi \frac{\partial \psi^*}{\partial t}\right) d\tau \tag{13.9}$$

For $\partial\psi/\partial t$ and $\partial\psi^*/\partial t$ we substitute the appropriate expressions from the time-dependent Schrödinger equation (Eq. 2.15) and its complex conjugate

$$\frac{\partial \psi}{\partial t} = \frac{i\hbar}{2m} \nabla^2\psi - \frac{i}{\hbar} V(r)\psi \tag{13.10}$$

and

$$\frac{\partial \psi^*}{\partial t} = -\frac{i\hbar}{2m} \nabla^2\psi^* + \frac{i}{\hbar} V(r)\psi^* \tag{13.11}$$

Thus Eq. 13.9 becomes

$$\frac{dP}{dt} = \frac{i\hbar}{2m} \int_V (\psi^* \nabla^2\psi - \psi \nabla^2\psi^*) \, d\tau \tag{13.12}$$

This can also be written as

$$\frac{dP}{dt} = \frac{i\hbar}{2m} \int_V \operatorname{div}(\psi^* \nabla\psi - \psi \nabla\psi^*) \, d\tau \tag{13.13}$$

[2] We had already ruled out a possible φ dependence.

Using the well-known divergence theorem linking the integral over the divergence of a vector with the surface integral over the vector, itself, Eq. 13.13 becomes

$$\frac{dP}{dt} = \frac{i\hbar}{2m} \int_S (\psi^* \, \nabla \psi - \psi \, \nabla \psi^*) \, d\mathbf{a} \tag{13.14}$$

where the integral is to be taken over the surface S bounding the volume V. If no particles are absorbed or created, and we excluded these possibilities, any change in the probability of finding a particle in the volume V must be connected with a particle flux[3] through the surface of the volume. The vector

$$\mathbf{S} = -\frac{i\hbar}{2m} (\psi^* \, \nabla \psi - \psi \, \nabla \psi^*) \tag{13.15}$$

must, therefore, represent a particle current, and there must be a continuity relation

$$\frac{dP}{dt} = -\int_V \text{div } \mathbf{S} \, d\tau = -\int_S \mathbf{S} \, d\mathbf{a} \tag{13.16}$$

The minus sign states, of course, that an outward flow—positive $\mathbf{S}$—causes a *decrease* in the probability of finding the particle inside the volume. Now we calculate the probability current for the incident plane wave

$$\psi = A' e^{ikz} \tag{13.17}$$

and get

$$\mathbf{S}' = -\frac{i\hbar}{2m} \{ A'^* e^{-ikz} A' ik e^{ikz} + A' e^{ikz} A'^* ik e^{-ikz} \}$$

$$= -\frac{i\hbar}{2m} 2 |A'|^2 \, ik = \frac{|A'|^2 \, \hbar k}{m} = |A'|^2 \, v = N' \tag{13.18}$$

since N' was defined earlier as the incident flux. For the scattered wave we obtain

$$\mathbf{S} = -\frac{i\hbar}{2m} \left\{ \frac{A^* f^*(\vartheta)}{r} e^{-ikr} A f(\vartheta) \frac{(ikr - 1)}{r^2} e^{ikr} \right.$$

$$\left. - \frac{A f(\vartheta)}{r} e^{ikr} A^* f^*(\vartheta) \frac{(-ikr - 1)}{r^2} e^{-ikr} \right\}$$

$$= \frac{\hbar |A|^2 |f|^2 \, k}{mr^2} = \frac{|A|^2 |f|^2 \, v}{r^2} \tag{13.19}$$

$\mathbf{S}$ is the flux through a surface element $d\mathbf{a}$. We assume the surface element to be orthogonal to the flux[4] and obtain the total number of particles penetrating

---

[3] Or, more precisely, a probability current.
[4] In other words, we assume that the detector is lined up properly.

$|d\mathbf{a}|$ by multiplying $|\mathbf{S}|$ with the surface element $|d\mathbf{a}| = r^2 \, d\Omega$. Thus, the number of particles scattered per second into $d\Omega$ is

$$N = |A|^2 \, |f|^2 \, v \, d\Omega \qquad (13.20)$$

We compare this with Eq. 13.1 and obtain

$$\sigma(E, \vartheta)N_n{}' = |A|^2 \, |f|^2 \, v \qquad (13.21)$$

The expression (Eq. 13.6) assumed, of course, scattering from one center. Hence $n = 1$. If no scattering takes place, the outgoing plane wave is equal to the incident wave; this determines the normalization constant:

$$A' = A \qquad (13.22)$$

Thus, according to Eq. 13.18, we have

$$\sigma(E, \vartheta) = |f(E, \vartheta)|^2 \qquad (13.23)$$

Still without committing ourselves to a particular potential V we can go one step further if we invoke our assumption that the potential has spherical symmetry.

In this case we can separate the Schrödinger equation just as we could for the bound particle. Since the angular part of the Schrödinger equation does not depend on either E or V (see Eq. 5.26) the solutions found in Chapter 5 must still be good. We excluded the possibility of a φ-dependence so that the solutions of the angular part of the Schrödinger equation must be Legendre polynomials. Thus, the most general solution of Eq. 13.3 becomes

$$u = \sum_{l=0} a_l \chi_l(r) P_l(\cos \vartheta) = \sum_{l=0} a_l u_l \qquad (13.24)$$

The radial functions $\chi_l(r)$ are, of course, no longer the functions we encountered in Chapter 5.

Equation 13.24 expresses the solutions of the Schrödinger equation (Eq. 13.3) as a superposition of **partial waves**.

$$u_l = \chi_l(r) P_l(\cos \vartheta) \qquad (13.25)$$

each belonging to one particular value of the angular momentum quantum number l. If, for instance, the distribution of the scattered particles is nearly independent of ϑ in the center-of-mass system, the dominant term in the expansion (Eq. 13.24) must be the one with $l = 0$, i.e $P_0(\cos \vartheta) = 1$; this case is often referred to as **s-wave scattering**.

13.3 PHASE SHIFT ANALYSIS

We shall now apply our method of partial wave expansion to a simple scattering problem. To be more explicit we shall apply it to the simplest scattering problem there is, $V = 0$ (or no scattering at all). Of course, we know that in this case the solution of the Schrödinger equation is a plane wave of the form

$$u = u_o e^{ikz} \tag{13.26}$$

since that was the form of the incident wave. If, however, we insist that this simple function be expanded according to Eq. 13.24, we shall have to solve the radial part of the Schrödinger equation in spherical polar coordinates for the case $V = 0$. This means that we must find a solution for the equation (see Eq. 5.25)

$$\frac{d}{dr}\left(r^2 \frac{d\chi(r)}{dr}\right) + \frac{2mr^2}{\hbar^2} E\chi(r) = l(l + 1)\chi(r) \tag{13.27}$$

To put this equation into a usable shape requires some mathematical legerdemain. First we introduce a new variable:

$$\rho = kr = \sqrt{\frac{2mE}{\hbar^2}}\, r$$

This yields

$$\frac{d}{d\rho}\left(\rho^2 \frac{d\chi(\rho)}{d\rho}\right) + \rho^2\chi(\rho) = l(l + 1)\chi(\rho) \tag{13.28}$$

next we substitute

$$\chi(\rho) = \rho^{-\frac{1}{2}}\eta(\rho) \tag{13.29}$$

This gives us

$$\rho^2 \frac{d\chi}{d\rho} = \rho^{\frac{3}{2}} \frac{d\eta}{d\rho} - \frac{\rho^{\frac{1}{2}}}{2}\eta \tag{13.30}$$

We substitute this into Eq. 13.27 and get after multiplication with $\rho^{\frac{1}{2}}$:

$$\rho^2 \frac{d^2\eta}{d\rho^2} + \rho \frac{d\eta}{d\rho} + [\rho^2 - l(l + 1) - \tfrac{1}{4}]\eta = 0 \tag{13.31}$$

This equation is known as Bessel's differential equation; its solutions are the Bessel functions[5]

$$\eta_l = J_v(\rho) = \left(\frac{\rho}{2}\right)^v \sum_{m=0}^{\infty} \frac{(-1)^m \left(\frac{\rho}{2}\right)^{2m}}{m!\,\Gamma(v + m + 1)} \tag{13.32}$$

where

$$v = \pm\sqrt{l(l + 1) + \tfrac{1}{4}} = \pm\tfrac{1}{2}(2l + 1) \tag{13.33}$$

[5] $\Gamma(x)$ is the well known *gamma function*; it satisfies $\Gamma(1) = 1$ and $\Gamma(x + 1) = x\Gamma(x)$. Consequently, $\Gamma(n + 1) = n!$ The gamma function is thus an extension of the factorial to noninteger and even negative numbers.

Thus, the complete solution of Eq. 13.31 is:

$$B_\nu(\rho) = aJ_\nu(\rho) + bJ_{-\nu}(\rho) \tag{13.34}$$

$J_{-\nu}$ diverges for $\rho \to 0$ and, since the probability amplitude should be finite everywhere, we conclude that the coefficient b in Eq. 13.34 must be zero.

Now we introduce a scattering potential, specifying that it be of short range or (to be more specific) that it satisfy:[6]

$$rV(r) \xrightarrow[r \to \infty]{} 0 \tag{13.35}$$

Since we have not specified the potential $V(r)$ in detail we cannot solve the inhomogeneous Bessel equation that now results for the radial part of the wave function. Fortunately we do not have to solve it, we are still interested only in the solutions in the region where the measurement is made, that is, where r is large. At large r—or ρ—, however, the scattered particles no longer feel the influence of the scattering potential. This means that we still have to solve the radial Schrödinger equation only for the asymptotic case $V = 0$, although for different boundary conditions at $r = 0$. Since the potential V is no longer zero *everywhere* the complete solution (Eq. 13.34) no longer applies in the region of small ρ. Since it does not describe the physical situation in this region we can no longer advance a physical reason that the value of the constant b in Eq. 13.34 be zero.

The asymptotic solution of the Schrödinger equation in the presence of a scattering potential thus becomes

$$u = \sum_{l=0} (a_l J_{l+\frac{1}{2}} + b_l J_{-l-\frac{1}{2}}) \frac{1}{\sqrt{\rho}} P_l(\cos \vartheta) \tag{13.36}$$

In view of the formidable nature of the Bessel functions it is desirable to find a—hopefully simplified—approximation for Eq. 13.32. As a starting point in this endeavor we use the well-known recurrence relation:[7]

$$\frac{d}{d\rho}\left(\frac{J_\nu(\rho)}{\rho^\nu}\right) = -\frac{J_{\nu+1}(\rho)}{\rho^\nu} \tag{13.37}$$

The index ν will, according to Eq. 13.33, assume only half-integer values. Thus we start with $J_{\frac{1}{2}}(\rho)$:

$$J_{\frac{1}{2}}(\rho) = \sqrt{\frac{\rho}{2}} \sum_{m=0}^{\infty} \frac{(-1)^m \rho^{2m}}{m! \, \Gamma(m + 1 + \frac{1}{2}) 2^{2m}} \tag{13.38}$$

Now

$$\Gamma(x + 1) = x\Gamma(x) \tag{13.39}$$

---

[6] Unless we insist on this condition (which is, incidentally, not met by the Coulomb potential $V(r) = q_1 q_2/r$), our formalism leads to unpleasant divergencies.

[7] See, for instance, A. E. Danese, *Advanced Calculus*, Allyn and Bacon Inc., Boston 1965, volume 1, p. 526.

Hence

$$\Gamma(x + n) = x(x + 1)(x + 2) \cdots (x + n - 1)\Gamma(x) \qquad (13.40)$$

Letting $x = \frac{1}{2}$ and $n = m + 1$ we get

$$J_{1/2}(\rho) = \sqrt{\frac{\rho}{2}} \sum_{m=0}^{\infty} \frac{(-1)^m \rho^{2m}}{m!\, 2^{2m} \cdot \frac{1}{2} \cdot \frac{3}{2} \cdot \frac{5}{2} \cdots (m + \frac{1}{2})\Gamma(\frac{1}{2})} \qquad (13.41)$$

Since

$$\Gamma(\tfrac{1}{2}) = \sqrt{\pi}$$

this is nothing but

$$J_{1/2}(\rho) = \sqrt{\frac{2}{\pi\rho}} \sum_{m=0}^{\infty} \frac{(-1)^m \rho^{2m+1}}{(2m + 1)!} = \sqrt{\frac{2}{\pi\rho}} \sin\rho \qquad (13.42)$$

Using the recurrence relation we can now obtain

$$J_{3/2}(\rho) = \sqrt{\frac{2}{\pi\rho}} \left(\frac{\sin\rho}{\rho} - \cos\rho \right) \qquad (13.43)$$

$$J_{5/2}(\rho) = \sqrt{\frac{2}{\pi\rho}} \left\{ \left(\frac{3}{\rho^2} - 1 \right) \sin\rho - \frac{3}{\rho} \cos\rho \right\} \qquad \text{etc.} \qquad (13.44)$$

At a large distance from the target we can neglect all the terms of higher order in ρ and get the following asymptotic expressions for the Bessel functions:

$$J_{1/2} \xrightarrow[\rho \to \infty]{} \sqrt{\frac{2}{\pi\rho}} \sin\rho \qquad (13.45)$$

$$J_{3/2} \xrightarrow[\rho \to \infty]{} -\sqrt{\frac{2}{\pi\rho}} \cos\rho \qquad (13.46)$$

$$J_{5/2} \xrightarrow[\rho \to \infty]{} -\sqrt{\frac{2}{\pi\rho}} \sin\rho \qquad (13.47)$$

$$J_{7/2} \xrightarrow[\rho \to \infty]{} \sqrt{\frac{2}{\pi\rho}} \cos\rho \qquad \text{etc.} \qquad (13.48)$$

In other words, except for the phase, all the Bessel functions of half-odd-integer order approach the same function. Since

$$\sin\left(\rho - \frac{\pi}{2} \right) = -\cos\rho, \qquad \sin(\rho - \pi) = -\sin\rho \qquad \text{etc.} \qquad (13.49)$$

we can write the asymptotic expressions for all the above Bessel functions in the form

$$J_{l+1/2} \xrightarrow[\rho \to \infty]{} \sqrt{\frac{2}{\pi\rho}} \sin\left(\rho - \frac{l\pi}{2} \right); \qquad l = 0, 1, 2, 3, \ldots \qquad (13.50)$$

In a completely analogous way we can show that

$$J_{-l-\frac{1}{2}} \xrightarrow[\rho \to \infty]{} (-1)^l \sqrt{\frac{2}{\pi \rho}} \cos \left(\rho - \frac{l\pi}{2}\right); \qquad l = 0, 1, 2, 3, \ldots \quad (13.51)$$

In the presence of a short-range scattering potential the asymptotic partial wave expansion of the wave function thus becomes:[8]

$$u = \sum_{l=0} \left[a_l \sin \left(kr - \frac{\pi l}{2}\right) + b_l' \cos \left(kr - \frac{\pi l}{2}\right) \right] \sqrt{\frac{2}{\pi}} \cdot \frac{1}{kr} P_l(\cos \vartheta) \quad (13.52)$$

In the absence of a scattering potential we found that the b_l' were zero. The b_l', therefore, must be some measure of the strength and range of the potential V and must be connected with the differential cross section $\sigma(E, \vartheta)$. Just how they are connected, we shall now try to find out. To this end we take a look at the lth partial wave.

$$u_l = \sqrt{\frac{2}{\pi}} \frac{1}{kr} \left\{ a_l \sin \left(kr - \frac{\pi l}{2}\right) + b_l' \cos \left(kr - \frac{\pi l}{2}\right) \right\} P_l(\cos \vartheta) \quad (13.53)$$

Using the identity

$$\sin (\alpha + \beta) = \sin \alpha \cos \beta + \cos \alpha \sin \beta \quad (13.54)$$

we can rewrite

$$a_l \sin \left(kr - \frac{\pi l}{2}\right) + b_l' \cos \left(kr - \frac{\pi l}{2}\right)$$

$$= c_l \sin \left(kr - \frac{\pi l}{2} + \delta_l\right)$$

$$= c_l \left[\sin \left(kr - \frac{\pi l}{2}\right) \cos \delta_l + \cos \left(kr - \frac{\pi l}{2}\right) \sin \delta_l \right] \quad (13.55)$$

Using this in Eq. 13.53, we can write the lth partial wave as

$$u_l = \sqrt{\frac{2}{\pi}} \frac{1}{kr} c_l \sin \left(kr - \frac{\pi l}{2} + \delta_l\right) P_l(\cos \vartheta) \quad (13.56)$$

If we compare the coefficients on both sides of Eq. 13.55, we find

$$\frac{b_l'}{a_l} = \tan \delta_l \quad (13.57)$$

The angle δ_l can, according to Eq. 13.56, be interpreted as a shift occurring in the phase of the lth partial wave. At this point we redefine our goal: It is

[8] We absorb the factor $(-1)^l$ in the b'.

customary to express the cross section in terms of these phase shifts δ_l rather than in terms of the coefficients b_l'. We have already established a connection between the scattering amplitude $f(\vartheta)$ and the cross section Eq. 13.23. It is, therefore, sufficient to relate the phase shifts δ_l to the scattering amplitude. In order to accomplish this we recall the two asymptotic expressions we found for the wave function (Eqs. 13.6 and 13.52) and equate them:

$$A\left\{e^{ikz} + \frac{1}{r}f(\vartheta)e^{ikr}\right\} = \sum_{l=0}^{\infty} c_l \frac{1}{kr}\sqrt{\frac{2}{\pi}} P_l(\cos \vartheta) \sin\left(kr - \frac{\pi l}{2} + \delta_l\right) \quad (13.58)$$

For large values of r this must be an identity in r as well as in ϑ. In the absence of a scattering potential this becomes:[9]

$$Ae^{ikz} = \sum_{l=0}^{\infty} a_l \frac{1}{kr}\sqrt{\frac{2}{\pi}} P_l(\cos \vartheta) \sin\left(kr - \frac{\pi l}{2}\right) \quad (13.59)$$

We substitute this expression on the left side of Eq. 13.58 and express the sines with the help of the identity $e^{i\alpha} = \cos \alpha + i \sin \alpha$. This yields

$$\sum_{l=0}^{\infty} \frac{a_l}{2ikr}\sqrt{\frac{2}{\pi}} P_l(\cos \vartheta)\{e^{ikr}e^{-i\pi l/2} - e^{-ikr}e^{i\pi l/2}\} + \frac{A}{r}f(\vartheta)e^{ikr}$$

$$= \sum_{l=0}^{\infty} c_l \frac{1}{2ikr}\sqrt{\frac{2}{\pi}} P_l(\cos \vartheta)\{e^{ikr}e^{-i\pi l/2}e^{i\delta_l} - e^{-ikr}e^{i\pi l/2}e^{-i\delta_l}\} \quad (13.60)$$

Since e^{ikr} and e^{-ikr} are linearly independent the coefficients of either must be equal on both sides of Eq. 13.60. We begin with the coefficients of e^{-ikr}:

$$-\sqrt{\frac{2}{\pi}}\frac{1}{2ikr}\sum_{l=0}^{\infty} a_l P_l(\cos \vartheta)e^{i\pi l/2} = -\sqrt{\frac{2}{\pi}}\frac{1}{2ikr}\sum_{l=0}^{\infty} c_l P_l(\cos \vartheta)e^{i\pi l/2}e^{-i\delta_l} \quad (13.61)$$

This equation holds for all values of ϑ and since $P_l = P_l(\cos \vartheta)$, the coefficients of P_l, must be equal on both sides, it follows that

$$c_l = a_l e^{i\delta_l} \quad (13.62)$$

Next we compare the coefficients of e^{ikr} in Eq. 13.60

$$\sqrt{\frac{2}{\pi}}\frac{1}{2ikr}\sum_{l=0}^{\infty} a_l P_l(\cos \vartheta)e^{-i\pi l/2} + \frac{A}{r}f(\vartheta) = \sqrt{\frac{2}{\pi}}\frac{1}{2ikr}\sum_{l=0}^{\infty} c_l P_l(\cos \vartheta)e^{-i\pi l/2}e^{i\delta_l}$$

$$(13.63)$$

We substitute Eq. 13.62 into Eq. 13.63 and solve for $f(\vartheta)$:

$$f(\vartheta) = \sqrt{\frac{2}{\pi}}\frac{1}{2ikA}\sum_{l=0}^{\infty} a_l P_l(\cos \vartheta)e^{-i\pi l/2}(e^{2i\delta_l} - 1) \quad (13.64)$$

[9] In the absence of a scattering potential we have, of course, $a_l = c_l$, $b_l = \delta_l = 0$.

The coefficients a_l in Eq. 13.64 are the coefficients that we obtained in Eq. (13.52) in the *absence* of a scattering potential. We, thus, can obtain them explicitly from the partial wave expansion of the plane wave Ae^{ikz}.

$$Ae^{ikr \cos \vartheta} = \sum_{n=0}^{\infty} a_n \frac{1}{kr} \sqrt{\frac{2}{\pi}} P_n(\cos \vartheta) \sin \left(kr - \frac{\pi n}{2}\right) \qquad (13.65)$$

We multiply Eq. 13.65 with $P_l(\cos \vartheta)$ and integrate

$$A \int_{-1}^{1} P_l(\cos \vartheta) e^{ikr \cos \vartheta} \, d \cos \vartheta$$

$$= \sum_{n=0}^{\infty} \frac{a_n}{kr} \sqrt{\frac{2}{\pi}} \sin \left(kr - \frac{\pi n}{2}\right) \int_{-1}^{1} P_l(\cos \vartheta) P_n(\cos \vartheta) \, d \cos \vartheta \qquad (13.66)$$

The right-hand side can be simplified by using the orthonormality relation between the Legendre polynomials:[10]

$$\int_{-1}^{1} P_l(\cos \vartheta) P_n(\cos \vartheta) \, d \cos \vartheta = \frac{2}{2l + 1} \delta_{nl} \qquad (13.67)$$

The left-hand side of Eq. 13.66 can be simplified by partial integration (using $\xi = \cos \vartheta$)

$$A \int_{-1}^{1} e^{ikr\xi} P_l(\xi) \, d\xi = \frac{A}{ikr} e^{ikr\xi} P_l(\xi) \Big|_{\xi=-1}^{\xi=1} - \frac{A}{ikr} \int_{-1}^{1} e^{ikr\xi} P'(\xi) \, d\xi \qquad (13.68)$$

Now $P_l(1) = 1$ and $P_l(-1) = (-1)^l$. Hence

$$\frac{A}{ikr} e^{ikr\xi} P_l(\xi) \Big|_{\xi=-1}^{\xi=1} = \begin{cases} \dfrac{A}{ikr} \, 2i \sin (kr) & \text{for even } l \\[2ex] \dfrac{A}{ikr} \, 2 \cos (kr) & \text{for odd } l \end{cases} \qquad (13.69)$$

Since

$$\sin \left(kr - \frac{\pi l}{2}\right) = \begin{cases} -\sin (kr) & \text{for even } l \\ -\cos (kr) & \text{for odd } l \end{cases} \qquad (13.70)$$

Equation 13.69 can also be written

$$\frac{A}{ikr} e^{ikr\xi} P_l(\xi) \Big|_{-1}^{1} = \frac{2Ai^l}{kr} \sin \left(kr - \frac{\pi l}{2}\right) \qquad (13.71)$$

The second term on the right-hand side of Eq. 13.68 can be shown to be small of second order in $1/r$ by partial integration. Now we substitute Eqs.

[10] As eigenfunctions of a hermitian operator the Legendre polynomials must be ortho-normal. The reader may verify the normalization expressed in Eq. 13.67.

13.71 and 13.67 into Eq. 13.66 and get

$$\frac{2Ai^l}{kr} \sin\left(kr - \frac{\pi l}{2}\right) = \sqrt{\frac{2}{\pi}} \frac{a_l}{kr} \sin\left(kr - \frac{\pi l}{2}\right) \frac{2}{2l+1} \tag{13.72}$$

or

$$a_l = Ai^l \sqrt{\frac{\pi}{2}} (2l+1) \tag{13.73}$$

If we substitute this into Eq. 13.64, we obtain

$$f(\vartheta) = \sum_{l=0}^{\infty} \frac{(2l+1)}{2ik} i^l P_l(\cos\vartheta) e^{-i\pi l/2}(e^{2i\delta_l} - 1) \tag{13.74}$$

Since

$$i^l = e^{i\pi l/2} \tag{13.75}$$

This can, finally, be written:

$$f(\vartheta) = \sum_{l=0}^{\infty} \frac{(2l+1)}{k} P_l(\cos\vartheta) e^{i\delta_l} \sin\delta_l \qquad \text{Phew!} \tag{13.76}$$

From 13.76 we can derive a simple expression for the total cross section $\sigma_o(E)$ by integrating $|f(\vartheta)|^2 = \sigma(E, \vartheta)$ over the solid angle (see Eqs. 13.2 and 13.23). To this end we separate the real and the imaginary part of $f(\vartheta)$:

$$f(\vartheta) = \sum_{l=0}^{\infty} \frac{(2l+1)}{k} P_l(\cos\vartheta) \cos\delta_l \sin\delta_l + i\sum_{l=0}^{\infty} \frac{(2l+1)}{k} P_l(\cos\vartheta) \sin^2\delta_l$$

$$\tag{13.77}$$

Hence:

$$\sigma_o(E) = \int_0^{2\pi} \int_0^{\pi} |f(\vartheta)|^2 \sin\vartheta \, d\vartheta \, d\varphi$$

$$= -2\pi \int_1^{-1} |f(\vartheta)|^2 \, d\cos\vartheta$$

$$= \frac{2\pi}{k^2} \int_{-1}^{1} \left[\sum_{l=0}^{\infty} (2l+1)P_l(\cos\vartheta)\right]^2 \cos^2\delta_l \sin^2\delta_l d\cos\vartheta$$

$$+ \frac{2\pi}{k^2} \int_{-1}^{1} \left[\sum_{l=0}^{\infty} (2l+1)P_l(\cos\vartheta)\right]^2 \sin^4\delta \, d\cos\vartheta \tag{13.78}$$

Using Eq. 13.67, this becomes

$$\sigma_o(E) = \frac{4\pi}{k^2} \sum_{l=0}^{\infty} (2l+1) \sin^2\delta_l \tag{13.79}$$

According to classical physics a particle passing a scattering center at a distance d with a velocity v has an angular momentum mvd. The quantum-mechanical angular momentum must for large l go over into the classical expression:

$$l(l+1)\hbar^2 \xrightarrow[l \to \infty]{} (mvd)^2 \tag{13.80}$$

Large values of l thus correspond to passage at a large distance d from the scattering center. We assumed the scattering potential to be of short range; thus we conclude that partial waves with large values of l have a small amplitude b_l'. Another way to express the same fact is to say: partial waves with large values of l have a small phase shift δ_l. The expansion (Eq. 13.79) should, thus, converge rapidly. This is most pronounced at low energies. Here v is small and, especially, if the scattering potential is of very short range, only the term with $l = 0$ contributes measurably. In this case (usually called s-wave scattering) the angular distribution of the scattered particles is isotropic in the center-of-mass system since $P(\cos \vartheta) = 1$. This wave-mechanical phenomenon, most prominently exhibited in low energy neutron scattering,[11] has a well-known optical analog: Objects that are small compared to the wavelength of the incident light scatter isotropically.

PROBLEMS

13.1 Draw the angular distribution of the particles for pure s-wave, p-wave, and d-wave scattering.

13.2 1 MeV neutrons are scattered on a target. The angular distribution of the neutrons in the center-of-mass system proves to be isotropic. The total cross section is measured to be 10^{-25} cm^2. Using the partial wave representation, calculate the phase shift of all the partial waves involved.

13.3 The total cross section for s-wave scattering on a short range potential cannot exceed a certain maximum value. Find this value and express it in terms of the de Broglie wavelength of the scattered particle.

13.4 Show that

$$J_{-l-\frac{1}{2}} \xrightarrow[\rho \to \infty]{} \sqrt{\frac{2}{\pi \rho}} \cos \left(\rho - \frac{\pi l}{2} \right); \qquad l = 0, 1, 2, 3, \ldots$$

13.5 A well-collimated beam of 1 MeV neutrons is shot through a target containing 10^{23} nuclei per cm^2. With a detector placed behind the target we find that only 83 percent of the neutrons penetrate the target. Attributing the loss to scattering, can we state whether a measurement of the angular distribution in the center-of-mass system would yield an isotropic distribution of the scattered neutrons? Explain your answer.

Solution of Eq. 13.5. 17 percent or 0.17 of all particles are scattered out of the beam by $n = 10^{23}$ nuclei. The total cross section of a target nucleus is thus given by

$$0.17 = \sigma_o \cdot 10^{23} \qquad \text{or} \qquad \sigma_o = 1.7 \cdot 10^{-24} \text{ cm}^2$$

[11] Why?

An isotropic distribution of the scattered particles would imply s-wave scattering. Equation 13.79 relates the total cross section to the angular momentum l and the sine of the phase shift δ_l. Pure s-wave scattering would mean that the sum breaks off after the first term, i.e.,

$$\sigma_o = \frac{4\pi}{k^2} \sin^2 \delta_l$$

now,

$$k = \frac{mv}{\hbar} = \frac{p}{\hbar} = \sqrt{\frac{2mT}{\hbar^2}}$$

where T is the kinetic energy of the neutrons. Substituting the proper values for m, T and $\hbar$ we obtain

$$k^2 = \frac{2 \cdot 1.67 \cdot 10^{-24} g \cdot 1.6 \cdot 10^{-6} \, erg}{(1.04 \cdot 10^{-27})^2 \, erg^2 \, sec^2} = 4.94 \cdot 10^{24} \, cm^{-2}$$

If our assumption of pure s-wave scattering were correct we would thus have

$$\sigma_o = 1.7 \cdot 10^{-24} = 2.54 \cdot 10^{-24} \cdot \sin^2 \delta_l$$

14

TO NEW FRONTIERS

In this chapter we shall outline, ever so briefly, some of the developments of quantum mechanics that go beyond the scope of this book.

14.1 RELATIVISTIC THEORIES

The famous **fine structure constant** $\alpha = v/c = 1/137$ is the ratio of the electron velocity in the first Bohr orbit and the velocity of light. The smallness of this number indicates that a nonrelativistic theory is sufficient for most atomic processes. It will fail if the particle velocity approaches the velocity of light (or to put it differently) if the kinetic energy of a particle approaches its rest energy:

$$E = m_o c^2$$

One obvious way to reconcile quantum mechanics with the special theory of relativity was immediately recognized by Schrödinger. Instead of making the transition to quantum mechanics from the classical expression[1]

$$E = \frac{mv^2}{2} \tag{14.1}$$

one can start from the relativistic equation[1]

$$E^2 = c^2 p^2 + m^2 c^4 \tag{14.2}$$

and substitute

$$E \rightarrow i\hbar \frac{\partial}{\partial t} \quad \text{and} \quad \mathbf{p} \rightarrow -i\hbar \nabla \tag{14.3}$$

The resulting wave equation

$$-\hbar^2 \frac{\partial^2 \psi}{\partial t^2} = -\hbar^2 c^2 \nabla^2 \psi + m^2 c^4 \psi \tag{14.4}$$

[1] For simplicity we restrict the discussion to the case of free particles ($V = 0$).

was soon discarded, since it did not describe the observed behavior of "particles" correctly. The reason for this is that the **"relativistic Schrödinger equation"** or **"Klein-Gordon equation"** Eq. 14.4 describes particles of integer spin (bosons) and such particles[2] had not been discovered at the time. The difficulties with the Klein-Gordon equation can be traced to the fact that it is of second order in the time derivative which appears as

$$-\hbar^2 \frac{\partial^2}{\partial t^2}$$

One can take the square root of the left side of Eq. 14.4 and obtain

$$i\hbar \frac{\partial}{\partial t}$$

but what is the meaning of the operator

$$\sqrt{-\hbar^2 c^2 \nabla^2 + m^2 c^4}$$

on the right-hand side? When **P. A. M. Dirac** attacked the same problem a few years later (1928), he boldly linearized Eq. 14.4 by writing

$$i\hbar \frac{\partial \psi}{\partial t} = \alpha i\hbar c \, \nabla \psi + \beta mc^2 \psi \tag{14.5}$$

Dirac found that this equation could be satisfied *if* β was a four-by-four matrix *and if* α was a three-component vector whose components were four-by-four matrices. The eigenfunctions ψ of Eq. 14.5 must then be four-component vectors, and it became soon apparent that two of these components were related to the existence of the two different spin states. The other two seemed to lead to states of negative energy for the free particle. This did not deter Dirac. He suggested that all the negative energy states in the world (allowed by the Pauli principle) might already be filled and that one would observe their existence only if an electron was knocked out of this infinite sea of negative-energy electrons into the regular world of positive energy. If this happened, a normal electron should appear together with a "hole" in the sea of negative energy electrons (see Figure 14.1). In the presence of an electric field pulling electrons, for example, to the right, the electron to the left of the hole would fill it, leaving a new hole to the left of the original one, etc. The hole in other words would behave like a particle with positive charge. Dirac thus postulated the existence of positrons four years before their actual discovery. Today the picture of the sea of negative energy particles is replaced by modern field theory which expands the basic

---

[2] An example is the π-meson.

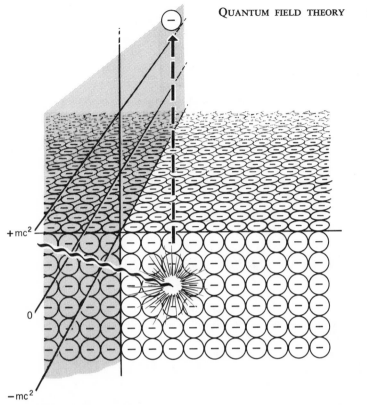

Fig. 14.1 A photon with an energy $h\nu \geqslant 2mc^2$ can lift an electron from a negative energy state to a state with positive energy. The remaining hole in the sea of negative energy states behaves in every way like a positive electron. (This kind of *pair production* requires the presence of a third particle—not shown here—in order to conserve energy and momentum.)

prediction of Dirac theory, saying that all particles have antiparticles. Particles and antiparticles have the same mass but opposite charge and magnetic moment. Neutral particles of integer spin can be identical with their antiparticles.

The description of spin with the help of Pauli matrices is, of course, a simplified form of the complete four-component theory of Dirac.

14.2 QUANTUM FIELD THEORY

We have "quantized" the equation of motion of classical mechanics and arrived at the Schrödinger equation. In a similar fashion we can quantize field equations. If these field equations describe the electromagnetic field, this leads to the formalism of quantum electrodynamics, which describes the emission and absorption of photons. Quantum electrodynamics, whose development was spurred by the discovery of the Lamb shift, is capable of making very precise predictions about the behavior of electrons and muons (predictions which thus far are in complete agreement with experiments). Among the lasting triumphs of quantum electrodynamics are the description of the Lamb shift and the exact prediction of the small deviation of the g-value of electrons and muons from the value given by the Dirac theory ($g = 2$). Quantization of the field of nuclear interactions leads to a field theory of mesons which, at this point, is much less developed than quantum electrodynamics.

A field theory of the weak interactions, modeled after quantum electrodynamics, has also been developed. Although this theory has not been as strikingly successful as quantum electrodynamics, the agreement between theory and experiment is usually quite good.

14.3 NEW QUANTUM NUMBERS: THE NUCLEON NUMBER, THE LEPTON NUMBER

Whenever physicists try to extend quantum mechanics to describe some new body of empirical evidence, they scan experimental results for manifestations of "quantum numbers" that are either conserved during interactions of the system or change by integer amounts. The next step is to find out if there are selection rules that govern the behavior of this quantum number in transitions from one state to another. An example may illustrate this. Neutrons and protons make up all known nuclei; they are very similar in mass, and the nuclear (strong) interaction between them seems to be independent of charge, i.e., another proton is just as strongly bound to a nucleus as another neutron.[3] These facts are taken into account by saying: Proton and neutron are two states of a particle called a nucleon. A neutron can decay into a proton, an electron, and a neutrino, but the free proton seems to be absolutely stable. For this reason one introduces the number of nucleons:

$$N = \text{(number of nucleons)} - \text{(number of antinucleons)} \qquad (14.6)$$

in a system as a quantum number and calls it the **nucleon number**. The nucleon number seems to be strictly conserved. True a neutron can decay into a proton, but a proton is still a nucleon and the lifetime of free protons (at least in our corner of the universe) is known to be larger than 10^{21} years.

Electrons are also known to be quite durable and one might think of introducing the "electron number": (number of electrons) − number of positrons) $= e$ as a new quantum number. It is, however, more practical to include the next of kin of the electron in the picture.

---

[3] Making due allowance for the fact that there exists an electrostatic repulsion between the proton and the other protons in the nucleus.

The muon is a very close relative of the electron; in fact, it is often called a heavy electron. It decays into an electron and two neutrinos. One calls this family of particles, the muon, the electron, and the neutrino:

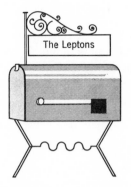

We claim the **lepton number** $L =$ (number of leptons) — (number of anti-leptons) is constant. Let us see how this works out:

A π-meson is no lepton. It decays into a muon and, if the lepton number is indeed conserved, its other decay product must be an antilepton. Since charge is always conserved the other decay product must be neutral and indeed we observe

$$\pi^+ \to \mu^+ + \nu \tag{14.7}$$

and

$$\pi^- \to \mu^- + \bar{\nu} \tag{14.8}$$

where $\bar{\nu}$ is the antineutrino. The muon, as stated above, decays into an electron (or positron) and two neutrinos. If we believe in lepton conservation, we must conclude that the two neutrinos are a neutrino and an antineutrino; hence

$$\mu^+ \to e^+ + \nu + \bar{\nu} \tag{14.9}$$

$$\mu^- \to e^- + \nu + \bar{\nu} \tag{14.10}$$

This was the story until 1964. At that time it was discovered that there are two different kinds of neutrinos—one ν_e associated with electrons and one ν_μ associated with muons. In the light of this new evidence we have to rewrite Eqs. 14.7 to 14.10 to read

$$\pi^+ \to \mu^+ + \nu_\mu \tag{14.11}$$

$$\pi^- \to \mu^- + \bar{\nu}_\mu \tag{14.12}$$

$$\mu^+ \to e^+ + \bar{\nu}_\mu + \nu_e \tag{14.13}$$

$$\mu^- \to e^- + \nu_\mu + \bar{\nu}_e \tag{14.14}$$

The reader may convince himself that Eqs. 14.11 to 14.14 indicate the conservation of two other quantum numbers sometimes—unfortunately— called *mu*-ness and *e*-ness.

14.4 ISOSPIN, STRANGENESS, HYPERCHARGE[4]

If neutron and proton are just two manifestations of the "nucleon," they must differ in at least one quantum number. To assume this quantum number to be the charge, is to underestimate theoretical physicists.

The existence of a particle "doublet" is reminiscent of the case of the electron which can also exist in two states:

"spin up"   or   "spin down"

Noting that proton and neutron are of (almost) equal mass, Heisenberg (1932) introduced the concept of "isotopic spin." Purists point out that the term isotopic spin is a misnomer since proton and neutron are not really isotopes, but differ in charge. For this reason the term isobaric spin has been proposed. We shall occupy a middle ground adopting the term **isospin** which seems to be gaining acceptance. In this new picture the nucleon has isospin $T = \frac{1}{2}$ and the third component T_3 (z-component) of the isospin vector points either up ($T_3 = \frac{1}{2}$) or down ($T_3 = -\frac{1}{2}$) in some isospin space. This enables one to bring the formalism of spin (Pauli matrices and all) to bear on the problem. The charge of a nucleon is given by

$$Q = T_3 + \frac{N}{2} \tag{14.15}$$

where N is the nucleon number. Obviously Eq. 14.15 leads to $Q = 0$ or $Q = +1$ for a single nucleon, depending on whether $T_3 = +\frac{1}{2}$ or $T_3 = -\frac{1}{2}$. All this may seem to be a complicated way of dealing with a simple situation. But there is more to it. In introducing the charge as the third component of an isospin vector we have also introduced the isospin, that is, the vector itself. Now, charge is conserved in all known reactions. This means the third component T_3 of the isospin vector is always conserved. However, the magnitude of the isospin vector, characterized by the quantum number T, is also conserved in those processes that are a result of the strong interactions.

The fact that these strong interactions are independent of the charge of the nucleon can thus be expressed by saying that strong interactions are independent of the orientation of the isospin vector in isospin space. This is

---

[4] All the considerations in this paragraph apply only to particles partaking in the strong interactions, not to leptons.

a close analogy to the case of the Coulomb part of the electromagnetic interaction, which is independent of the orientation of the spin vector in real space.

If we want to be sophisticated, we can sum this up as follows: Proton and neutron, the two isospin states of the nucleon, are degenerate. The electromagnetic interaction lifts the degeneracy and leads to the small mass difference between the two particles. Presumably this is more than just a nice analogy but, without a detailed understanding of the strong interactions, who can tell.

The concept of isospin can be extended to other strongly interacting particles, and we apply Eq. 14.15 to the case of the π-meson. The π-meson is no nucleon. Hence $N = 0$. The negative pion has a charge $Q = -1$ which means that T_3 must be equal to -1. If $T_3 = -1$, T, at least, must be equal to 1. If $T = 1$ and if our isospin formalism is worth anything, there must be components $T_3 = 1, 0, -1$. In other words, there should be three different π-mesons with very similar masses forming an isospin triplet. The charge of the three π-mesons must be $Q = 1$, $Q = 0$, and $Q = -1$. This is exactly what one observes.

Applied to another well-known meson, the K-meson (Eq. 14.15), breaks down completely. The K-meson is no nucleon. Hence $N = 0$. Its charge is ± 1 or zero, but there are two distinctly different neutral K-mesons. This indicates that the K-mesons do not form an isospin triplet with $T = 1$ but that the K^+ and the K° mesons are members of an *isospin doublet* and that the $\overline{K^+} = K^-$ and the $\overline{K^\circ}$ are their antiparticles. If this is so we must have $T = \frac{1}{2}$ and $T_3 = \pm\frac{1}{2}$. Unwilling to give up Eq. 14.15 completely, M. Gell-Mann modified it to read

$$Q = T_3 + \frac{N + S}{2} \qquad (14.16)$$

introducing a new quantum number S. He named this quantum number **strangeness** for reasons not germane to this discussion. The introduction of strangeness resolves another puzzle concerning the behavior of K-mesons. These particles have the peculiar property of being readily produced in high-energy nucleon-nucleon collisions but decay only very reluctantly by way of weak interactions. This seems to violate the general principle stated on p. 152 that a large probability for the transition

state 1 → state 2

implies a large probability for the transition

state 2 → state 1

But strangeness is more than just a farfetched attempt to salvage Eq. 14.15. Strong interactions conserve strangeness and since nucleons have strangeness

$S = 0$, K-mesons and other strange particles can only be produced in pairs of opposite strangeness in nucleon-nucleon collisions (associated production). The K-meson is the lightest strange particle and cannot decay by strong interactions (i.e., with a large probability or short lifetime). Once formed it has to sit around until the weak interactions, which do not worry about strangeness, cause it to decay.

N and S are often lumped together in one quantum number

$$Y = N + S \tag{14.17}$$

which is called **hypercharge.** By this definition pions have no hypercharge. Kaons and nucleons have hypercharge $Y = 1$. After many futile attempts to bring order into the chaos of the elementary particles, which were discovered by the score every year, it seems now that the diligent search for new quantum numbers is beginning to pay off. Some wondrous symmetries based on the two quantum numbers T and Y are beginning to emerge, promising for the first time that an understanding of this brave new world of elementary particles may not be far off.

MATHEMATICAL REVIEW

A.1 LINEAR OPERATORS

It is often convenient to use an operator notation in writing a differential equation. An operator is an instruction to carry out a certain mathematical operation on the symbol following it.

Examples

$$\frac{\partial}{\partial x}(\cdots); \quad \nabla(\cdots); \quad \nabla^2(\cdots), \quad \text{etc.}$$

The advantage of considering mathematical instructions as operators lies in the fact that operators often lead to algebraic identities regardless of the nature of the function they are applied to and, thus, make it simpler to carry out the instructions they contain.

Examples

$$\frac{\partial}{\partial x} \cdot \frac{\partial}{\partial y} - \frac{\partial}{\partial y} \cdot \frac{\partial}{\partial x} \equiv 0 \quad \text{(for analytic functions)} \tag{A.1}$$

$$x\frac{\partial}{\partial x} - \frac{\partial}{\partial x}x \equiv -1 \tag{A.2}$$

or written explicitly:

$$x\frac{\partial f(x, y\cdots)}{\partial x} - \frac{\partial}{\partial x}[xf(x, y\cdots)]$$

$$\equiv x\frac{\partial f(x, y\cdots)}{\partial x} - f(x, y, \ldots) - x\frac{\partial f(x, y, \cdots)}{\partial x}$$

$$\equiv -f(x, y, \ldots) \tag{A.2a}$$

In the following we shall restrict ourselves to the consideration of **linear operators,** a class of operators that is of special interest to physicists.

Definition.   An operator A that satisfies

$$A(\psi + \varphi) = A\psi + A\varphi \tag{A.3}$$

and

$$A\lambda\psi = \lambda A\psi \tag{A.4}$$

(where λ is a constant) is said to be a linear operator. The reader may convince himself that the above examples all satisfy the criteria (Eqs. A.3 and A.4). Examples of nonlinear operators would be:

$$\sin(\cdots) \qquad \text{or} \qquad \cos(\cdots)$$

which obviously do not satisfy Eq. A.3 or Eq. A.4. An operator A, applied to a function $\psi(\mathbf{r}, t)$ will usually change it into a completely different function, i.e.,

$$A\psi(\mathbf{r}, t) = \varphi(\mathbf{r}, t) \tag{A.5}$$

There are, however, certain functions which, if certain operators are applied to them, are merely multiplied by a (possibly complex) constant, i.e.,

$$Au(\mathbf{r}, t) = au(\mathbf{r}, t) \tag{A.6}$$

where a is a constant. Such functions are called eigenfunctions[1] of the operator; the constant a is called the eigenvalue[1] of the eigenfunction (with respect to a particular operator).

Example. e^{ix} is an eigenfunction of the operator d/dx; its eigenvalue is i. The same function is also an eigenfunction to the operator $-2id/dx$ with the eigenvalue 2. As linear operator H is called **hermitian** if for two functions ψ and φ

$$\int \psi^* H\varphi \, d\tau = \int (H\psi)^* \varphi \, d\tau \tag{A.7}$$

The property of being hermitian is called **hermiticity**.

Example.   The Laplace operator is hermitian.

Proof

$$\nabla(\varphi\nabla\psi^*) = \nabla\varphi\nabla\psi^* + \varphi\nabla^2\psi^* \tag{A.8}$$

$$\nabla(\psi^*\nabla\varphi) = \nabla\psi^*\nabla\varphi + \psi^*\nabla^2\varphi \tag{A.9}$$

Subtracting on both sides and integrating yields:

$$\int \nabla(\varphi \, \nabla\psi^* - \psi^* \, \nabla\varphi) \, d\tau = \int \varphi \, \nabla^2\psi^* \, d\tau - \int \psi^* \, \nabla^2\varphi \, d\tau \tag{A.10}$$

---

[1] This is the terminology used by physicists. Mathematicians usually prefer the terms, characteristic functions and characteristic numbers.

On the left-hand side we have an integral over the divergence of the vector $(\varphi\nabla\psi^* - \psi^*\nabla\varphi)$. We remember Gauss's theorem:

$$\int \text{div (vector)}\, d\tau = \int \text{vector} \cdot d\mathbf{A} \tag{A.11}$$

where $d\mathbf{A}$ is a surface element and the integration on the right goes over the surface of the integration volume on the left. Hence:

$$\int \nabla(\varphi\,\nabla\psi^* - \psi^*\,\nabla\varphi)\, d\tau = \int (\varphi\,\nabla\psi^* - \psi^*\,\nabla\varphi) \cdot d\mathbf{A} = 0 \tag{A.12}$$

if we assume that for large values of r both ψ and φ go sufficiently fast to zero to make the integral vanish. It will turn out that the functions that we deal with in this book either will always vanish sufficiently fast to satisfy Eq. A.12 or can be so selected that they vanish on some closed surface. We return to Eq. A.10 which, considering Eq. A.12 yields

$$\int \varphi\,\nabla^2\psi^*\, d\tau = \int \psi^*\,\nabla^2\varphi\, d\tau \qquad \text{q.e.d.} \tag{A.13}$$

It should be emphasized that the proof of the hermiticity of ∇^2 was based on the vanishing of the surface integral (Eq. A.12). This shows that the hermiticity of an operator depends on the nature of the functions to which it is applied. As stated above we will always be justified in assuming that integrals of the type in Eq. A.12 vanish.

A.2 THE FOURIER SERIES

A well-known mathematical device for dealing with otherwise unmanageable functions is to express them as an infinite Taylor series. This enables one, for instance, to integrate functions to any desired degree of approximation even if direct analytic integration is not possible. Actually, the Taylor series is only one of infinitely many ways to express a function as an infinite sum, and we shall now investigate a series expansion first derived by Fourier (1822) that is much more useful to the physicist.

Let $f(t)$ be a function which is periodic with the period τ but otherwise not unduly restricted. In particular, it does not have to have a finite derivative everywhere (as would be required for the expansion into a Taylor series), nor does it need to be more than piecewise continuous, i.e., it can have a finite number of jumps. A plot of the function could, for instance, look like any of the curves in Figure A.1.

Fourier showed that it was possible to expand any such function into an infinite series of sines and cosines of multiples of the fundamental frequency: $\omega = 2\pi/\tau$

$$f(t) = \sum_0^\infty a_k \cos(k\omega t) + \sum_1^\infty b_k \sin(k\omega t) \tag{A.14}$$

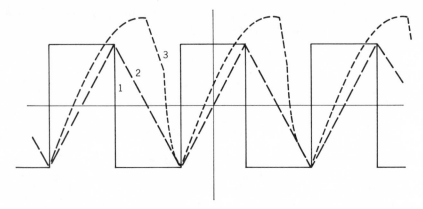

Fig. A.1 Any periodic function can be represented by a Fourier series.

The advantage of such a Fourier decomposition (or Fourier analysis) in physical applications is obvious: Very frequently it is easy to measure or calculate the response of a system (for example, a radio or an amplifier) to monochromatic sine waves. If the system is linear, i.e., if the output is proportional to the input (at a given frequency), Eq. A.14 allows one to calculate its response to any periodic input if the response to sine waves of various frequencies is known.

To determine the constant coefficients a_k and b_k we write

$$f(t) = \sum_0^n a_k \cos{(k\omega t)} + \sum_1^n b_k \sin{(k\omega t)} + \epsilon_n(t) \qquad (A.15)$$

Obviously any function $f(t)$ can be written in this form since we have introduced the new function

$$\epsilon_n(t)$$

to make up for any difference between $f(t)$ and the finite sums

$$\sum_0^n a_k \cos{(k\omega t)} + \sum_1^n b_k \sin{(k\omega t)} \qquad (A.16)$$

If expansion (Eq. A.14) is to be possible, $\epsilon_n(t)$ must vanish in some suitably defined way[2] as n goes to infinity. We shall now show that $|\epsilon_n(t)|^2$ can be minimized through a suitable choice of the coefficients A_k and B_k. We form the mean square deviation:

$$M = \frac{1}{\tau} \int_{-\tau/2}^{\tau/2} [\epsilon_n(\xi)]^2 \, d\xi = \frac{1}{\tau} \int_{-\tau/2}^{\tau/2} \left[f(\xi) - \sum_0^n a_k \cos{(k\omega\xi)} - \sum_1^n b_k \sin{(k\omega\xi)} \right]^2 d\xi$$

$$(A.17)$$

[2] The reason for hedging a little will soon become apparent.

We minimize M by requiring

$$\frac{\partial M}{\partial a_k} = \frac{\partial M}{\partial b_k} = 0 \qquad \text{for} \qquad k = 0, 1, 2, \ldots, n \qquad (A.18)$$

This yields $n + 1$ equations:

$$\frac{\partial M}{\partial a_l} = \frac{1}{\tau} \int_{-\tau/2}^{\tau/2} 2\left[f(\xi) - \sum_0^n a_k \cos(k\omega\xi) - \sum_1^n b_k \sin(k\omega\xi) \right] \cos(l\omega\xi) \, d\xi = 0$$

$$(A.19)$$

and n similar equations $\partial M/\partial b_l = 0$. Now

$$\frac{2}{\tau} \int_{-\tau/2}^{\tau/2} \cos(l\omega\xi) \cos(k\omega\xi) \, d\xi = \delta_{lk} \begin{cases} = 1 & \text{if} \quad l = k \\ = 0 & \text{if} \quad l \neq k \end{cases} \qquad (A.20)$$

also

$$\frac{2}{\tau} \int_{-\tau/2}^{\tau/2} \cos(l\omega\xi) \sin(k\omega\xi) \, d\xi = 0 \qquad (A.21)$$

Hence from Eqs. A.19, A.20 and A.21

$$\frac{2}{\tau} \int_{-\tau/2}^{\tau/2} f(\xi) \cos(k\omega\xi) \, d\xi = a_k \qquad (A.22)$$

$$\frac{2}{\tau} \int_{-\tau/2}^{\tau/2} f(\xi) \sin(k\omega\xi) \, d\xi = b_k \qquad (A.23)$$

and:

$$\frac{1}{\tau} \int_{-\tau/2}^{\tau/2} f(\xi) \, d\xi = a_o \qquad (A.24)$$

The fact that Eqs. A.22, A.23, and A.24 no longer depend on n means that the same a_k and b_k that minimize M for one value of n minimize it for any value of n. In other words, the a_k and b_k determined by Eqs. A.22, A.23, and A.24 are final.

For many applications it is convenient to write Eq. A.14 in complex form. To this end we substitute Eqs. A.22 to A.24 into Eq. A.14

$$f(t) = \frac{1}{\tau} \int_{-\tau/2}^{\tau/2} f(\xi) \, d\xi + \frac{2}{\tau} \sum_1^n \int_{-\tau/2}^{\tau/2} f(\xi) \cos(k\omega\xi) \cos(k\omega t) \, d\xi$$

$$+ \frac{2}{\tau} \sum_1^n \int_{-\tau/2}^{\tau/2} f(\xi) \sin(k\omega\xi) \sin(k\omega t) \, d\xi + \epsilon_n(t) \quad (A.25)$$

(since $\sin(k\omega t)$ and $\cos(k\omega t)$ do not contain the integration parameter ξ we can include them under the integral). Using the identity

$$\cos(\alpha - \beta) = \cos\alpha \cdot \cos\beta + \sin\alpha \cdot \sin\beta \qquad (A.26)$$

this becomes

$$f(t) = \frac{1}{\tau} \int_{-\tau/2}^{\tau/2} f(\xi) \, d\xi + \frac{2}{\tau} \sum_{1}^{n} \int_{-\tau/2}^{\tau/2} f(\xi) \cos [k\omega(t - \xi)] \, d\xi + \epsilon_n(t) \quad \text{(A.27)}$$

Using the identity

$$\cos \alpha = \tfrac{1}{2}(e^{i\alpha} + e^{-i\alpha}) \quad \text{(A.28)}$$

we obtain

$$f(t) = \frac{1}{\tau} \int_{-\tau/2}^{\tau/2} f(\xi) \, d\xi + \frac{1}{\tau} \sum_{1}^{n} \int_{-\tau/2}^{\tau/2} f(\xi)[e^{ik\omega(t-\xi)} + e^{-ik\omega(t-\xi)}] \, d\xi + \epsilon_n(t)$$

$$= \frac{1}{\tau} \sum_{-n}^{n} \int_{-\tau/2}^{\tau/2} f(\xi) e^{ik\omega(t-\xi)} \, d\xi \quad \text{(A.29)}$$

With the abbreviations

$$c_k = \frac{1}{\tau} \int_{-\tau/2}^{\tau/2} f(\xi) e^{-ik\omega\xi} \, d\xi \quad \text{(A.30)}$$

$$c_{-k} = c_k{}^* \quad \text{(A.31)}$$

$$c_o = a_0 \quad \text{(A.32)}$$

we can finally write

$$f(t) = \sum_{-n}^{n} c_k e^{ik\omega t} + \epsilon_n(t) \quad \text{(A.33)}$$

For the actual convergence proof we refer to one of the standard works on analysis.[3] If the function to be expanded has discontinuities (and we have not excluded this possibility), the convergence of its Fourier expansion does not imply that

$$\lim_{n \to \infty} \sum_{-n}^{n} c_k e^{ik\omega t} = f(t) \quad \text{(A.34)}$$

but rather that the weaker statement

$$\lim_{n \to \infty} \left[\sum_{k=-n}^{n} c_k e^{ik\omega t} - f(t) \right]^2 = 0 \quad \text{(A.35)}$$

holds true. Whereas Eq. A.34 would mean that $f(t)$ could be approximated everywhere with arbitrary precision, Eq. A.35 states only that the *average deviation* between $f(t)$ and its Fourier series vanishes.

This allows for the possibility that local discrepancies exist. Such discrepancies occur only in the vicinity of discontinuities of $f(t)$ and are known as Gibbs' phenomenon. Figure A.2 shows several finite Fourier series approximating a square wave. The over- and undershoot in the vicinity of

[3] For example, Whittaker and Watson, *Modern Analysis*, Cambridge University Press, 1963, p. 174 f.f.

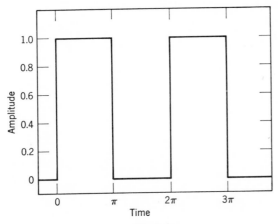

Fig. A.2*a*

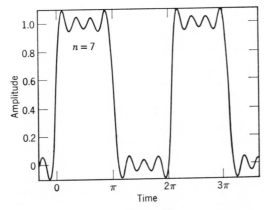

Fig. A.2*b*

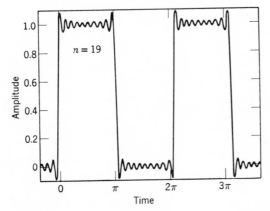

Fig. A.2*c*

257

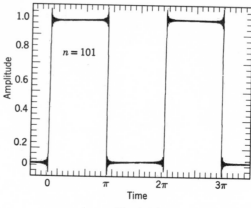

Fig. A.2*d*

Fig. A.2*a,b,c,d* A square wave (Fig. A.2*a*) can be approximated with a Fourier series. Shown here are Fourier series breaking off at $n = 7$ (Fig. A.2*b*), $n = 19$ (Fig. A.2*c*), and $n = 101$ (Fig. A.2*d*). Notice how Gibb's phenomenon gradually evolves at the discontinuities.

the discontinuity does not vanish as one goes to larger and larger values of n, but rather is compressed more and more towards the discontinuity. Its maximum amplitude approximates in this case ± 18 percent for very large n.

The Fourier expansion of a periodic function is of great importance in many areas of physics. The Fourier analyzer best known to the experimental physicist is, of course, the spectrometer. It takes a periodic nonsinusoidal (light) wave and expands it into a series of sinusoidal waves of different frequency (spectral lines)[4]

$$\sin (k\omega t)$$

and intensity

$$|a_k|^2$$

The fact that the coefficients are given in the form of integrals may deter the novice (after all integration is more difficult than differentiation). Actually, it is an advantage in practical applications. Integrations can more easily be performed by a computer, and it is much simpler and much more accurate to integrate a measured curve than it is to differentiate it. The determination of the Fourier coefficients (Eqs. A.22 to A.24 and Eqs. A.30 and A.32) was facilitated by the existence of the relations (Eqs. A.20 and A.21) that made the coefficients final, i.e., independent of n. As it turns out, the trigonometric

[4] A spectrometer does not usually determine the phase and, therefore, allows no distinction between $\sin (k\omega t)$ and $\cos (k\omega t)$.

functions are not the only ones to have relations of this kind and, hence, they are not the only functions that can be used for series expansions in the spirit of Eq. A.33. This subject is explored in detail in Chapter 6.1.

A.3   THE FOURIER INTEGRAL

The intriguing possibilities offered by the Fourier series make one wonder whether it is possible to use a similar expansion for non periodic functions. A non periodic function is one whose period is infinite.

In the preceding paragraph we said nothing about the size τ of the period, and we investigate now what happens if we let τ go to infinity. We substitute

$$\omega_k = k\omega = \frac{2\pi k}{\tau} \tag{A.36}$$

and

$$\Delta\omega_k = (k+1)\omega - k\omega = \omega = \frac{2\pi}{\tau} \tag{A.37}$$

into Eq. A.29:

$$f(t) = \sum_{-\infty}^{\infty} \frac{\Delta\omega_k}{2\pi} \int_{-\tau/2}^{\tau/2} f(\xi)e^{i\omega k(t-\xi)} \, d\xi \tag{A.38}$$

If we let τ grow, $\Delta\omega_k = 2\pi/\tau$ becomes smaller and smaller, ω_k becomes a continuous variable, and the sum (Eq. A.38) goes over into the integral.

$$f(t) = \frac{1}{2\pi} \int_{-\infty}^{\infty} d\omega \int_{-\infty}^{\infty} f(\xi)e^{i\omega(t-\xi)} \, d\xi \tag{A.39}$$

The second integration can be carried out, and the integral[5]

$$A(\omega) = \frac{1}{\sqrt{2\pi}} \int_{-\infty}^{\infty} f(\xi)e^{-i\omega\xi} \, d\xi \tag{A.40}$$

is a function of the frequency ω. We rewrite Eq. A.39 in the form[5]:

$$f(t) = \frac{1}{\sqrt{2\pi}} \int_{-\infty}^{\infty} A(\omega)e^{i\omega t} \, d\omega \tag{A.41}$$

If we compare Eq. A.41 with A.33, the meaning of $A(\omega)$ becomes apparent: The Fourier series expresses a periodic function in terms of sines and cosines[6] whose frequencies were multiples of the *fundamental frequency* ω

[5] To emphasize the symmetry between the Fourier transform and its inverse, it is customary to split the factor $1/2\pi$ in Eq. A.39.
[6] Sines and cosines are lumped together in the complex exponential; negative frequencies just change the sign (phase) of the sine term.

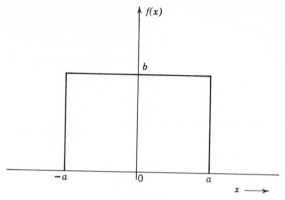

Fig. A.3

and whose amplitudes varied discretely with k. The Fourier integral expresses a non periodic function as a superposition of sines and cosines[6] of continuously varying frequency whose *amplitude $A(\omega)$ varies continuously* with ω.

Example. Find the *Fourier integral*, also called **Fourier transform,** of the following simple function (see Figure A.3):

$$f(x) = b \quad \text{for} \quad -a \leqslant x \leqslant a$$
$$f(x) = 0 \quad \text{for} \quad -a > x > a$$

For convenience we make the area under the "curve" equal to unity, i.e., $2ab = 1$.

Solution

$$A(\omega) = \frac{1}{\sqrt{2\pi}} \int_{-a}^{a} b e^{-i\omega\xi}\, d\xi = \frac{-ib}{\sqrt{2\pi}\,\omega}\left(e^{i\omega a} - e^{-i\omega a}\right) = \sqrt{\frac{2}{\pi}}\,\frac{b}{\omega}\sin\left(\omega a\right) \quad \text{(A.42)}$$

The *frequency spectrum $A(\omega)$* of the discontinuous function of Figure A.3 extends to higher and higher frequencies as a decreases.[7] For an infinitely short "pulse" we obtain

$$\lim_{a \to 0} A(\omega) = \sqrt{\frac{2}{\pi}}\,\frac{ab\omega}{\omega} = \frac{1}{\sqrt{2\pi}}$$

The amplitude has become independent of the frequency, i.e., the spectrum extends to infinity.

---

[7] Keeping, of course, $2ab = 1$.

A.4   MATRICES

Let us consider a vector **y** in an n-dimensional space. If we are given a coordinate system in this space, we can express the vector in terms of its projections on the n different coordinate axes and obtain its components $y_1, y_2, \ldots, y_n$ with respect to this coordinate system. We can write these components in the form of a column

$$\mathbf{y} = \begin{pmatrix} y_1 \\ y_2 \\ \cdot \\ \cdot \\ \cdot \\ y_n \end{pmatrix} \tag{A.43}$$

and call this column the representative of the vector in the given coordinate system. We can change the vector **y** into another vector **z** by means of a transformation A

$$A\mathbf{y} = \mathbf{z} \tag{A.44}$$

A transformation is an operation which can change both the direction and the length (magnitude) of the vector. In the following we shall consider only **linear transformations.**

Definition.   A transformation A is said to be linear if for any (possibly complex) constant λ

$$A(\lambda\mathbf{x}) = \lambda A\mathbf{x} \tag{A.45}$$

and if for any two vectors **x** and **y**

$$A(\mathbf{x} + \mathbf{y}) = A\mathbf{x} + A\mathbf{y} \tag{A.46}$$

Since in the most general of all linear transformations any component of the representative of **y** may depend on any component of the representative of **x**, the former can be expressed in terms of the latter by writing

$$y_1 = \sum_1^n a_{1l}x_l$$

$$y_2 = \sum_1^n a_{2l}x_l \tag{A.47}$$

or in general

$$y_k = \sum a_{kl}x_l$$

[8] A very convenient and frequently used convention is to write

$$\sum a_{kl}x_l = a_{kl}x_l$$

with the understanding that one has to sum over the index that appears twice (here l). However, in this book we shall, adhere to the more explicit notation of Eq. A.47.

The constant coefficients a_{kl} can be arranged in a two-dimensional not necessarily quadratic scheme which is called a matrix.

$$A = \begin{pmatrix} a_{11} & a_{12} & \cdots & a_{1n} \\ a_{21} & a_{22} & \cdots & \cdot \\ \cdot & \cdot & & \cdot \\ \cdot & \cdot & & \cdot \\ \cdot & \cdot & & \cdot \\ a_{n1} & & & a_{nn} \end{pmatrix} \tag{A.48}$$

The matrix A obviously has a similar relationship to the transformation A as the column in Eq. A.43 has to the vector y. A is the representative of the transformation in our particular coordinate system.

Now we ask, what happens if x itself has been created from a vector z by another linear transformation B, i.e., if

$$x_l = \sum_{m=1}^{n} b_{lm} z_m \tag{A.49}$$

In this case the connection between the y_k and the z_m must be given by

$$y_k = \sum_{l=1}^{n} a_{kl} \left(\sum_{m=1}^{n} b_{lm} z_m \right) = \sum_{m=1}^{n} \left(\sum_{l=1}^{n} a_{kl} b_{lm} \right) z_m = \sum_{m=1}^{n} c_{km} z_m \tag{A.50}$$

The c_{km} also form a matrix with the elements:

$$c_{km} = \sum_{l=1}^{n} a_{kl} b_{lm} \tag{A.51}$$

Comparing the three matrices

$$A = \begin{pmatrix} a_{11} & a_{12} & \cdots & a_{1n} \\ \cdot & & & \\ \cdot & & & \\ \cdot & & & \\ a_{k1} & a_{k2} & \cdots & a_{kn} \\ \cdot & & & \\ \cdot & & & \\ \cdot & & & \end{pmatrix};$$

$$B = \begin{pmatrix} b_{11} & \cdots & b_{1m} & \cdots & b_{1n} \\ b_{21} & \cdots & b_{2m} & \cdots & b_{2n} \\ \cdot & & \cdot & & \\ \cdot & & \cdot & & \\ \cdot & & \cdot & & \end{pmatrix}; \quad C = \begin{pmatrix} c_{11} & \cdots & c_{1m} & \cdots & c_{1n} \\ \cdot & & \cdot & & \\ \cdot & & \cdot & & \\ \cdot & & \cdot & & \\ c_{k1} & \cdots & c_{km} & \cdots \\ \cdot & & \cdot & & \\ \cdot & & \cdot & & \end{pmatrix}$$

$$\tag{A.52}$$

we realize that the element c_{km} is obtained by successively multiplying the first, second, etc., coefficient of the kth row of $\mathbf{A}$ with the first, second, etc., coefficient of the mth column of $\mathbf{B}$ and adding all the products. The result $\mathbf{C}$ is called the product of the two matrices $\mathbf{A}$ and $\mathbf{B}$, written

$$\mathbf{A} \cdot \mathbf{B} = \mathbf{C} \tag{A.53}$$

From Eq. A.51 follows that matrices do not in general commute, i.e., $\mathbf{A} \cdot \mathbf{B} \neq \mathbf{B} \cdot \mathbf{A}$; hence, we must distinguish between multiplication from the left and multiplication from the right. The sum $\mathbf{A} + \mathbf{B} = \mathbf{D}$ of two matrices is defined by

$$a_{ik} + b_{ik} = d_{ik} \tag{A.54}$$

A matrix $\tilde{\mathbf{A}}$, obtained from a matrix $\mathbf{A}$ by interchanging rows and columns, is said to be the **transpose** of $\mathbf{A}$, written in terms of the coefficients:

$$\tilde{a}_{ik} = a_{ki} \tag{A.55}$$

Obviously: $\tilde{\tilde{\mathbf{A}}} = \mathbf{A}$. In quantum-mechanical applications we frequently encounter matrices whose coefficients are complex, and the complex conjugate $\mathbf{A}^*$ of a matrix $\mathbf{A}$ is obtained by taking the complex conjugate of each coefficient.

Definition.   The **hermitian adjoint** $\mathbf{A}\dagger$ of a matrix $\mathbf{A}$ is obtained by taking the complex conjugate of the transpose of $\mathbf{A}$.

$$\mathbf{A}\dagger = \tilde{\mathbf{A}}^* \tag{A.56}$$

Definition.   A matrix that is equal to its hermitian adjoint

$$\mathbf{A}\dagger = \mathbf{A} \tag{A.57}$$

is said to be **hermitian** (or selfadjoint)

Definition.   The unit matrix $\mathbf{1}$ is a matrix whose elements are 1 in the main diagonal and zero everywhere else, i.e.,

$$a_{kk} = 1 \qquad a_{ik} = 0 \qquad \text{for} \qquad i \neq k$$

or

$$a_{ik} = \delta_{ik}$$

where δ_{ik} is the **Kronecker symbol,** which is defined as being 1 for $i = k$ and 0 for $i \neq k$.

Definition.   A matrix $\mathbf{A}^{-1}$ for which holds

$$\mathbf{A}^{-1} \cdot \mathbf{A} = \mathbf{A} \cdot \mathbf{A}^{-1} = \mathbf{1} \tag{A.58}$$

is said to be the inverse of $\mathbf{A}$. The inverse $\mathbf{A}^{-1}$ of a matrix $\mathbf{A}$ has the elements[9]

$$a_{ik}^{-1} = \frac{\alpha_{ki}}{\det \mathbf{A}}^{\,10} \tag{A.59}$$

[9] See Problem A.11.
[10] Note: α_{ki} *not* α_{ik}.

det $\mathbf{A}$ is the determinant we obtain if we consider the elements of $\mathbf{A}$ the elements of a determinant. Obviously we must insist that $\mathbf{A}$ is a square matrix and that det $\mathbf{A} \neq 0$ in order for the inverse to be defined. α_{ki} is the cofactor of the element a_{ki} and is defined as

$$\alpha_{ki} = \frac{\partial}{\partial a_{ki}} (\det \mathbf{A}) \tag{A.60}$$

The reader may convince himself that α_{ki} is the determinant that we obtain if we set the kth row and the ith column of det $\mathbf{A}$ equal to zero. We can now define another kind of matrix that is of great importance to quantum mechanics.

Definition. A matrix $\mathbf{U}$ is said to be unitary if it satisfies

$$\mathbf{U}^\dagger\mathbf{U} = \mathbf{U}\mathbf{U}^\dagger = \mathbf{1} \quad \text{or} \quad \mathbf{U}^\dagger = \mathbf{U}^{-1} \tag{A.61}$$

Definition. A vector $\mathbf{x}$ for which holds

$$\mathbf{A}\mathbf{x} = \lambda\mathbf{x} \tag{A.62}$$

where λ is a, possibly complex, constant, is said to be an eigenvector to the matrix $\mathbf{A}$. λ is called the eigenvalue of $\mathbf{x}$ with respect to $\mathbf{A}$.

PROBLEMS

A.1 Write down the explicit expressions for ∇r, ∇r^2, $\nabla(1/r)$, and $\nabla^2(1/r)$, in cartesian coordinates.

A.2 Show that the operator ∇ is not hermitian. Can this operator be made hermitian by multiplication with a constant?

A.3 Show that a hermitian operator is always linear.

A.4 Derive the Fourier series that describes the function plotted in Figure A.4.

A.5 Can the function $y = \sin x$ be expressed as a series

$$y = \sum_{k=0}^{\infty} a_k \cos (kx)$$

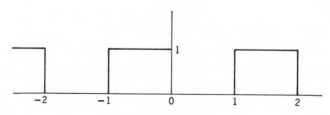

Fig. A.4

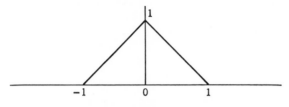

Fig. A.5

A.6   What is the Fourier transform of the function plotted in Figure A.5.

A.7   What is the complex conjugate of the adjoint of a hermitian matrix?

A.8   Show that

$$\widetilde{(AB)} = \tilde{B}\tilde{A}$$

A.9   Show that

$$(ABC)\dagger = C\dagger B\dagger A\dagger$$

A.10 Can a matrix be both unitary and hermitian? Must a unitray matrix be hermitian?

A.11 Verify Eq. A.59.

A.12 Show that the application of a unitary transformation to a vector leaves its magnitude unchanged.

SOLUTIONS

A.3   The definition of a linear operator is given in Eqs. A.45 and A.46:

$$A(\lambda x) = \lambda A x \quad \text{and} \quad A(x + y) = Ax + Ay$$

Let H be a hermitian operator and λ a constant, then

$$(x, H\lambda y) = (Hx, \lambda y) = \lambda(Hx, y) = \lambda(x, Hy) = (x, \lambda Hy)$$

or

$$H\lambda y = \lambda H y$$

for any y. Furthermore

$$(z, H(x + y)] = (Hz, x + y) = (Hz, x) + (H, z, y) = (z, Hx) + (z, Hy)$$
$$= (z, Hx + Hy)$$

or

$$H(x + y) = Hx + Hy$$

A.12 Let U be the unitary transformation that transforms x into x'

$$x' = Ux$$

If the length (magnitude of x remains invariant under this transformation, we must have

$$x' \cdot x' = x \cdot x$$

In the following we must keep track carefully of the ordering of the various terms. This task will be greatly eased if we consider the vectors $\mathbf{x}$ and $\mathbf{x}'$ as one-row or one-column matrices. The dot product of two vectors becomes in this case, using the conventions of matrix multiplication, the multiplication of a one-row matrix with a one-column matrix. If

$$\mathbf{x} = \begin{pmatrix} x_1 \\ x_2 \\ x_3 \end{pmatrix}$$

is the column form, then the dot product should be written as

$$\tilde{\mathbf{x}}\mathbf{x} = x_1{}^2 + x_2{}^2 + x_3{}^2$$

If we want to allow for the possibility that x is complex, we have to write

$$\mathbf{x}\dagger\mathbf{x} = x_1{}^*x_1 + x_2{}^*x_2 + x_3{}^*x_3$$

With this possibility in mind we thus write

$$\mathbf{x}'\dagger x' = (U\mathbf{x})\dagger U\mathbf{x} = \mathbf{x}\dagger U\dagger U\mathbf{x} = \mathbf{x}\dagger\mathbf{x} \qquad \text{q.e.d.}$$

INDEX

كاظم جواد كاظم